WITHDRAWN

Forensic Issues in Alcohol Testing

Forensic Issues in Alcohol Testing

Edited by
Steven B. Karch, MD, FFFLM

Consultant Pathologist and Toxicologist
Berkeley, California

CRC Press is an imprint of the
Taylor & Francis Group, an informa business

CRC Press
Taylor & Francis Group
6000 Broken Sound Parkway NW, Suite 300
Boca Raton, FL 33487-2742

Printed in the United States of America on acid-free paper
10 9 8 7 6 5 4 3 2 1

International Standard Book Number-13: 978-1-4200-5445-3 (Hardcover)

Library of Congress Cataloging-in-Publication Data

Forensic issues in alcohol testing / [edited by] Steven B. Karch.
p. ; cm.
"A CRC title."
Includes bibliographical references and index.
ISBN-13: 978-1-4200-5445-3 (hardcover : alk. paper)
ISBN-10: 1-4200-5445-7 (hardcover : alk. paper)
1. Blood alcohol--Analysis. 2. Breath tests. 3. Drug testing. 4. Automobile drivers--Alcohol use. I. Karch, Steven B.
[DNLM: 1. Alcoholic Intoxication--diagnosis. 2. Ethanol--adverse effects. 3. Ethanol--analysis. 4. Forensic Medicine--methods. 5. Forensic Toxicology--methods. W 775 F7154 2007] I. Title.

RA565F6779 2007
615'.7828--dc22 2007008112

Visit the Taylor & Francis Web site at
http://www.taylorandfrancis.com

and the CRC Press Web site at
http://www.crcpress.com

Contents

Preface

This book provides an overview of clinical and forensic-medical aspects of man's favorite recreational drug, alcohol (ethanol or ethyl alcohol). Consumption of alcoholic beverages is increasing worldwide as are many of the undesirable consequences of heavy drinking and drunkenness. The effects produced by alcohol depend on the amounts consumed and the speed of drinking. Small quantities cause euphoria and feelings of well-being whereas large amounts and high blood-alcohol concentration (BAC) depress the central nervous system and cause a decrement in performance and behavior, especially where skilled tasks like driving are concerned.

The various chapters of this book reflect different areas of interest for both forensic scientists and clinical and forensic toxicologists. The acute effects of alcohol are major concerns for motor and cognitive functioning. This is important for traffic safety because alcohol intoxication and drunkenness are incriminated in many fatal crashes. Enforcement of drinking and driving laws throughout the world depends to a large extent on the concentration of alcohol measured in a specimen of blood, breath, or urine obtained from the suspect. This kind of "chemical testing" to produce evidence for prosecution requires the use of highly reliable analytical methods to guarantee legal security for the individual. The widespread adoption of concentration per se laws for driving under the influence of alcohol tends to create a razor-sharp difference in penalty for those close to the statutory limit. Small analytical or pre-analytical errors might make the difference in penalty for those close to the statutory limit. Small analytical or pre-analytical errors might make the difference between acquittal or punishment in borderline cases.

In addition to details of the analytical methods used to measure alcohol in body fluids, knowledge is also provided about the disposition and fate of alcohol in the body and the factors influencing absorption, distribution, and elimination processes. Alcohol also tops the list of drugs encountered in postmortem toxicology. Although the methods of analysis are the same as for living subjects, making a correct interpretation of the results is often problematic. When dealing with autopsy specimens, various artifacts can arise because of the poor quality of blood and body fluid specimens, postmortem diffusion and redistribution, sampling site differences, and the risk of postmortem synthesis owing to microbial activity.

Judging whether a person drinks too much alcohol is not always easy because many people deny that they have a drinking problem. Tolerance is also an issue. Heavy drinkers become tolerant to many of alcohol's effects and do not manifest the same degree of impairment as would be seen in the naive drinker. This complicates making an early diagnosis and commencement of treatment for those at greatest risk of becoming dependent on alcohol. Clearly there is an urgent need to develop more objective ways to verify overconsumption of alcohol. Considerable research effort has been devoted to evaluate biochemical tests with enough sensitivity and specificity to detect hazardous drinking before this escalates to the point of causing organ and tissue damage. The final chapter of this book describes recent advances in the field of alcohol biomarkers that are intended to disclose both acute and chronic consumption of alcohol as well as relapse after a period of rehabilitation.

The Editor

Steven B. Karch, M.D., FFFLM, received his undergraduate degree from Brown University. He attended graduate school in anatomy and cell biology at Stanford University. He received his medical degree from Tulane University School of Medicine. Dr. Karch did postgraduate training in neuropathology at the Royal London Hospital and in cardiac pathology at Stanford University. For many years he was a consultant cardiac pathologist to San Francisco's Chief Medical Examiner.

In the U.K., Dr. Karch served as a consultant to the Crown and helped prepare the cases against serial murderer Dr. Harold Shipman, who was subsequently convicted of murdering 248 of his patients. He has testified on drug abuse–related matters in courts around the world. He has a special interest in cases of alleged euthanasia, and in episodes where mothers are accused of murdering their children by the transference of drugs, either *in utero* or by breast feeding.

Dr. Karch is the author of nearly 100 papers and book chapters, most of which are concerned with the effects of drug abuse on the heart. He has published seven books. He is currently completing the fourth edition of *Pathology of Drug Abuse*, a widely used textbook. He is also working on a popular history of Napoleon and his doctors.

Dr. Karch is forensic science editor for Humana Press, and he serves on the editorial boards of the *Journal of Cardiovascular Toxicology*, the *Journal of Clinical Forensic Medicine* (London), *Forensic Science, Medicine and Pathology*, and *Clarke's Analysis of Drugs and Poisons*.

Dr. Karch was elected a fellow of the Faculty of Legal and Forensic Medicine, Royal College of Physicians (London) in 2006. He is also a fellow of the American Academy of Forensic Sciences, the Society of Forensic Toxicologists (SOFT), the National Association of Medical Examiners (NAME), the Royal Society of Medicine in London, and the Forensic Science Society of the U.K. He is a member of The International Association of Forensic Toxicologists (TIAFT).

Contributors

Anders Helander, Ph.D.
Department of Clinical Neuroscience
Karolinska Institute and Karolinska University Hospital
Stockholm, Sweden

Alan Wayne Jones, D.Sc.
Department of Forensic Toxicology
University Hospital
Linköping, Sweden

Christopher S. Martin, Ph.D.
Western Psychiatric Institute and Clinic
Department of Psychiatry
University of Pittsburgh School of Medicine
Pittsburgh, Pennsylvania

Derrick J. Pounder, M.D.
Department of Forensic Medicine
University of Dundee
Scotland, U.K.

J. Robert Zettl, B.S., M.P.A., DABFE
Forensic Consultants, Inc.
Centennial, Colorado

CHAPTER 1

Measuring Acute Alcohol Impairment

Christopher S. Martin, Ph.D.
Western Psychiatric Institute and Clinic, Department of Psychiatry, University of Pittsburgh School of Medicine, Pittsburgh, Pennsylvania

CONTENTS

1.1 INTRODUCTION

This chapter reviews impairment produced by alcohol consumption, and issues in the measurement of such impairment. Motor and cognitive impairment produced by alcohol intoxication has been noted for centuries, and is apparent to almost everyone who lives in an alcohol drinking society. The effects of alcohol consumption on behavior, cognition, and mood have been reviewed by several authors in the scientific literature.[1-4] Unlike previous reviews, this chapter focuses on medical and forensic aspects of the topic, with a particular emphasis on impairment testing.

Impairment and its consequences are major reasons for forensic and medical interest in alcohol consumption. The assessment of impairment caused by acute alcohol intoxication is important for forensic, research, and clinical applications. At the same time, the determinants of impaired performance are complex, and are influenced by numerous pharmacological, motivational, and situational factors. Impairment also is extremely variable between persons. While obvious at extreme levels, alcohol-related impairment raises a number of difficult measurement issues. Although many impairment tests have adequate properties in the laboratory, assessments used in field applications have important limitations.

This chapter begins with a review of the acute effects of alcohol on behavioral and cognitive functioning. Emphasis is given to the effects of alcohol on speech and on the functioning of the vestibular system, which is centrally involved in balance control and spatial orientation. Next, individual differences in impairment are examined. This is followed by a description of how impairment is related to the time course of alcohol ingestion. The effects of rising vs. falling blood alcohol concentrations (BACs) and acute tolerance are discussed. The behavioral correlates of a hangover are reviewed. Then we discuss the ideal characteristics of impairment tests, followed by a description of the actual characteristics of impairment tests when evaluated in laboratory and field settings.

1.2 BEHAVIORAL CORRELATES OF ACUTE INTOXICATION

This chapter reviews the effects of alcohol on motor coordination and cognitive performance. Most "behavioral" correlates of intoxication involve both motor control and cognitive functioning. As will be seen, the effects of alcohol are not uniform; impairment varies across different types of behavioral functions. Two areas of functioning that are sensitive to alcohol impairment and assessed in field sobriety tests are described in some detail: speech and vestibular functioning. In addition, this chapter describes individual differences in alcohol impairment related to age, gender, and alcohol consumption practices.

1.2.1 Motor Control and Cognitive Functioning

Alcohol functions as a general central nervous system depressant, and affects a wide range of functions. To the observer, one of the most apparent effects of alcohol consumption is on motor control, particularly behaviors that require fine motor coordination. Other well-known effects of alcohol involve decrements in the cognitive control of behavioral functioning, especially the ability to perform and coordinate multiple tasks at the same time. Research has assessed the effects of alcohol on numerous performance tasks. Almost all of these tasks involve both cognitive and motor control components, although tasks differ in the complexity of the motor and cognitive functioning required for performance.

1.2.1.1 Reaction Time

Some of the most basic performance tasks investigated in the literature are simple reaction time (RT) tasks, in which subjects must push a button as quickly and accurately as possible in response to a stimulus. Baylor et al.[5] found no effects of alcohol on simple RT at BACs of 100 mg/dL, but did find effects at very high BACs near 170 mg/dL. Taberner[6] found no effect of a low dose of 0.15 g/kg alcohol, and a small effect at a dose of 0.76 g/kg alcohol. Maylor et al.[7] found a small effect of alcohol on simple RT at a dose of 0.64 g/kg alcohol. Linnoila et al.[8] did not find effects of alcohol on simple RT, even though other types of performance were impaired at this dose. Although the findings are somewhat variable, simple RT appears to be relatively insensitive to alcohol consumption.

Other research has examined choice RT tasks, in which subjects are required to respond using two or more buttons in response to different stimuli. Such tasks involve motor speed as well as cognitive functions involved in categorizing a stimulus and choosing a response. Few differences were seen by Fagan et al.[9] on a six-choice RT procedure. Golby[10] found no effects of alcohol on two different choice RT tasks. Other studies have found rather small or inconsistent effects of alcohol on choice RT.[7,11] Overall, there is not consistent evidence that alcohol affects the performance of choice RT tasks in the range of BACs studied in laboratory experiments.

The research literature contains a number of studies that have examined the effects of alcohol on tracking performance, that is, the ability of subjects to move a pointer to track a moving target. Tracking tasks require fine motor control and coordination of the hands and eyes at a rapid speed. There is consistent evidence that alcohol significantly impairs tracking performance. Beirness and Vogel-Sprott[12] found that alcohol affected tracking performance at BACs above 50 mg/dL; this effect has been replicated in numerous studies by Vogel-Sprott and colleagues. Wilson et al.[13] found that BACs peaking at 100 mg/dL produced tracking impairments for 120 min after drinking relative to placebo. The effects of alcohol on tracking performance have been found by several other authors.[8,11,14,15]

1.2.1.2 Dual-Task Performance

Alcohol impairment is found consistently during dual-task performance, when subjects are required to perform multiple tasks simultaneously. When subjects were required to perform a tracking task and an RT task at the same time, Connors and Maisto[11] found that alcohol reduced tracking but not RT performance. Using a similar procedure, Maylor et al.[14] found that alcohol affected RT but not tracking. Differences between these two studies in the task that was affected may have been due to instructions or task demands that led subjects to select one of the two tasks as primary, leading to performance deficits on the secondary task. Niaura et al.[16] combined an RT task with a task requiring the subject to circle target characters on a printed sheet, and found that alcohol produced deficits on both tasks relative to placebo. Other researchers have used computerized divided-attention tasks, which require subjects to perform multiple functions simultaneously. Mills and Bisgrove[17] found that divided-attention performance (responding to numbers on central and peripheral monitors by pushing different buttons) was impaired after 0.76 g/kg alcohol, but not after a lower dose of 0.37 g/kg alcohol.

The performance of multiple challenging tasks is thought to require the utilization of a large amount of attention, defined as limited-capacity cognitive resources that are required for effortful processing. The demonstration that alcohol produces dual-task performance deficits is consistent with the idea that alcohol produces impairment, in part, by reducing the available amount of limited-capacity attentional resources. A similar "attention allocation" model was proposed by Steele and Josephs.[18] These authors found that alcohol produced clear deficits in secondary task performance without affecting primary task performance, and suggested that alcohol serves to allocate a greater amount of attentional resources to a primary task, leading to fewer resources available for processing secondary sources of information.

Other studies of the behavioral impairment produced by alcohol have used tests designed to simulate complex real-world behaviors such as driving. Accident statistics consistently demonstrate that crash risk in the natural environment increases significantly when BACs are above 40 mg/dL.[19] Automobile simulator studies generally find that the information processing and lateral guidance demands of driving are adversely affected by alcohol. Several well-designed laboratory studies have demonstrated adverse affects of alcohol on skills related to driving, beginning with BACs as low as 30 to 40 mg/dL.[15,20] Other research with automobile simulators has examined the effects of alcohol on risk taking, defined by levels of speed, cars overtaken, and number of accidents during simulated driving. McMillen et al.[21] did not find effects of alcohol on risk taking during a driving simulation, whereas Mongrain and Standing[22] did find that alcohol increased risk taking, albeit at very high BACs near 160 mg/dL.

The effects of alcohol on performance have also been studied under conditions of actual driving. Attwood et al.[23] found that performance variables such as velocity and lane position together discriminated between intoxicated and sober drivers. Huntley found decreased lateral guidance during a driving task after alcohol.[24] Brewer and Sandow[25] studied real accidents using driver and witness testimony. Among persons who were in accidents, those with BACs above 50 mg/dL were much more likely to have been engaged in a secondary activity at the time of the accident, compared to drivers with BACs below 50 mg/dL. Overall, it appears that alcohol adversely affects several types of behavioral functions involved in driving. Driving performance is complex and determined by a number of individual and situational factors and roadway conditions. More research is needed to better understand alcohol's effects on driving performance.

1.2.2 Speech

It is well known to bartenders, law enforcement personnel, and the general public that alcohol consumption can produce changes in speech production often described as "slurred speech." Because speech production requires fine motor control, timing, and coordination of the lips, tongue, and vocal cords, it may be a sensitive index of impairment resulting from alcohol intoxication. Having subjects recite the alphabet at a fast rate of speed is a well-known field sobriety test. Laboratory research suggests that speech can be a valid index of alcohol consumption. After consuming 10 oz of 86-proof alcohol, alcoholics were found to take longer to read a passage and had more word, phrase, and sound interjections, word omissions, word revisions, and broken suffixes in their speech.[26] Other research with nonalcoholic drinkers found that under intoxication, subjects made more sentence-level, word-level, and sound-level errors during spontaneous speech.[27,28] Intoxicated talkers consistently lengthen some speech sounds, particularly consonants in unstressed syllables.[29] The overall rate of speech also slows when intoxicated talkers read sentences and paragraphs.[27,30]

Pisoni and Martin[30] examined the acoustic-phonetic properties of speech for matched pairs of sentences spoken by social drinkers when sober and after achieving BACs above 100 mg/dL. Sentence duration was increased after drinking, and pitch (loudness), while not consistently higher or lower, was more variable. The strongest effects of alcohol at the sound level were for speech sounds that require fine motor control and timing of articulation events in close temporal proximity. Intoxicated talkers displayed difficulty in controlling the abrupt closures and openings of the vocal tract required for stops and affricate closures. This resulted in long durations of closures before voiced stops (e.g., /d/, /b/), and the complete absence of closures before affricates (e.g., the /ch/ in "church"). These effects are consistent with what is known about the degree of precision of motor control mechanisms required for the articulation of different speech sounds. Pisoni and Martin[30] also found that listeners can reliably discriminate speech produced while sober and under intoxication. State Troopers showed higher discrimination levels than other listeners, suggesting that experience in detecting intoxication may increase perceptual abilities. The approach of some field sobriety tests that have persons recite the alphabet quickly may effectively capture the detrimental effects of alcohol on the articulation of speech sounds in close temporal proximity.

Despite the data showing effects of alcohol on speech, there are a number of limitations that make it difficult to use speech production as an index of alcohol impairment. Changes in speech have been reliably produced with blood alcohol levels above 100 mg/dL; however, the effects of lower doses have been variable;[27–29] it is not clear whether reliable effects are produced in most persons when BACs are lower. Other types of impairment are likely to occur before speech is noticeably affected. It is not clear from the literature how motivation to avoid the detection of intoxication would affect speech production. Furthermore, the specificity of speech changes to alcohol intoxication (rather than fatigue, stress, etc.) needs further study. Finally, there is extreme variability between persons in the acoustic-phonetic properties of speech, such that it is difficult to estimate the degree of impairment without comparison samples of sober speech.[31] Despite these limitations, the literature suggests that speech is likely to be a good screening test for impairment, which can then be determined using other measures.

1.2.3 Vestibular Functioning

The vestibular system serves to maintain spatial orientation and balance, and eye movements that support these functions. The vestibular system is comprised of two sets of interconnected canals that provide information about spatial orientation. Each canal is comprised of a membrane embedded with sensory hair cells, and a surrounding extracellular fluid. The otolithic canals provide information about the direction of gravity relative to the head, and thus are sensitive to lateral (side to side) head movements. This is accomplished by the fact that the membrane has a specific gravity that is twice that of the extracellular space in otoliths. Under normal conditions, the semicircular canals are sensitive to rotational movements of the head, and do not respond to lateral movements. In semicircular canals, the membrane and the extracellular fluid have the same exact specific gravity (i.e., weight by volume), such that the hair cells have neutral buoyancy and are not subject to gravitational influences.[32]

1.2.3.1 Positional Alcohol Nystagmus (PAN)

Alcohol's effects on the vestibular system are seen in measures that evaluate oculo-motor control, i.e., the functional effectiveness of eye movements under different conditions. During alcohol consumption, many persons show significant nystagmus (jerkiness) in eye movements when the head is placed in a sideways position: this effect is known as Positional Alcohol Nystagmus (PAN). There are two types of PAN. PAN I is characterized by a nystagmus to the right when the right side of the head is down, and to the left when the left side of the head is down. PAN I normally occurs during rising and peak BACs, beginning around 40 mg/dL.[33] PAN II normally appears between 5 and 10 hours after drinking, and is characterized by nystagmus in the opposite directions seen in PAN I.[34]

The mechanisms of PAN I and II have been convincingly demonstrated.[35] Both types of PAN are produced by the effects of alcohol on the semicircular canals, making hair cells on the membrane responsive to the effects of gravity. As alcohol diffuses throughout the water compartments of the body, it first enters the membrane space (which is richly supplied with capillary blood), and diffuses only gradually into the extracellular fluid. For a time, the alcohol concentration is greater in membrane than the surrounding fluid. Because alcohol is lighter than water, the specific gravity of the membrane will be lighter than that of the surrounding fluid during this time, making the semicircular canals responsive to gravity and producing PAN I. The faster the rate of drinking, the faster PAN I appears.[36]

There is a period during descending BACs in which neither PAN I nor PAN II is apparent; during this time the semicircular membrane and the surrounding fluid have achieved equilibrium and have the same specific gravity. PAN II occurs during alcohol elimination and after the body has no measurable amounts of alcohol. During PAN II, alcohol in the semicircular canals is removed from the membrane faster than the surrounding fluid; this results in the membrane having a heavier specific gravity than the surrounding fluid, which in turn produces PAN II.

Both PAN I and II may overstimulate the semicircular canals in a manner similar to motion sickness.[35] It is possible that the effects of alcohol on the semicircular canals play a central role in many symptoms of intoxication, including feelings of dizziness, nausea, and the experience of vertigo known as the "bedspins." Laboratory studies have shown that the magnitude of PAN I is associated with higher BACs,[36] with greater impairment in postural control, and with higher subjective intoxication ratings.[37] As described later, it has been speculated that PAN II is associated with hangover.[35]

1.2.3.2 Horizontal Gaze Nystagmus (HGN)

Another type of nystagmus produced by alcohol is known as Horizontal Gaze Nystagmus (HGN). HGN is defined by jerkiness in eye movements as gaze is directed to the side, when the head is in an upright position. HGN has long been noted as an effect of alcohol, and usually becomes apparent when rising BACs reach about 80 mg/dL.[38] HGN is assessed by having a subject follow an object with his eyes at an increasingly eccentric angle, without moving the head; the smallest angle at which nystagmus first appears is used to assess intoxication. While nystagmus occurs in a sober state at more extreme angles of eccentric gaze, alcohol decreases the size of the angle at which it is first apparent.

As with PAN, it has been demonstrated that HGN is highly associated with the effects of alcohol, as seen in studies of oculo-motor control.[39] Lehti[40] reported high correlations of BACs and the angle at which nystagmus in eccentric gaze became apparent. Similar effects of alcohol have been seen when nystagmus is assessed during active and passive head movements.[41] Tharp et al.[42] quantified the slope of a regression line predicting angle of onset of nystagmus from BAC. The angle of nystagmus onset was predicted at 45° for BACs of 50 mg/dL; 40° for BACs of 100 mg/dL; and 35° for BACs of 150 mg/dL. The angle of horizontal nystagmus onset has been found to have a high level of sensitivity and specificity in predicting BACs above 100 mg/dL in an emergency room setting.[43] HGN appears to be pharmacologically specific to alcohol. This is not the case for other aspects of occular control such as smooth pursuit. Smooth pursuit eye movements have proven to be much more sensitive to alcohol compared with marijuana, which has very small effects.[44] However, smooth pursuit eye movements are significantly affected by benzodiazapines, barbiturates, and antihistamines,[45] and thus cannot be said to be pharmcologically specific to alcohol. Deficits in smooth pursuit in the absence of significant BACs may indicate that a person has taken sedative drugs.

Postural control tasks are some of the most widely used measures of alcohol-related impairment in the laboratory and in the field. It is likely that the functional effectiveness of the vestibular system is an important locus of the effects of alcohol on postural control. Numerous studies have demonstrated that alcohol consumption leads to increases in sway as measured on a variety of balance platforms and similar types of apparatuses, appearing in many drinkers at BACs of 30 to 50 mg/dL.[9,13,46,47] Other research has shown that sway increases with alcohol dose,[17,48,49] and that heavier drinkers sway less after alcohol compared to lighter drinkers.[48] The effects of alcohol become greater as the postural control task becomes more difficult, such as with eyes closed, or when the feet are in a heel-to-toe position.[49] Thus, postural control appears to be a sensitive index of alcohol effects. However, body sway shows a great deal of individual variation in a sober condition. For this reason, the ability to detect impairment from measuring sway is limited in field settings in which sober performance measures are not available.

1.2.4 Individual Differences

There are large individual differences in the impairment produced by alcohol consumption. The first and most obvious difference is that persons differ in the BACs they achieve when drinking alcohol. Even when controlling for BAC, however, there are large differences between persons in their sensitivity to alcohol impairment. Perhaps the most important factor is drinking practices. Those who drink more often and in greater amounts tend to develop a greater amount of tolerance to the impairing effects of alcohol, i.e., have an acquired decrease in the degree of impairment across multiple drinking sessions. Greater impairment in light drinkers compared to heavier drinkers has been shown by numerous authors using a variety of performance tasks.[4,48,50]

There are also gender differences in some aspects of alcohol pharmacokinetics. On average, women achieve significantly higher BACs than men when drinking the same amount of alcohol because of mean gender differences in body weight and body fat,[50] and because females tend to have lower levels of gastric alcohol dehydrogenase.[51] Some research with women has found greater alcohol elimination rates[52] and greater sensitivity to alcohol effects[53,54] during the mid-luteal and ovulatory phases of the menstrual cycle compared to the follicular phase. Other research, however, has not replicated these findings.[16,55,56]

Some laboratory studies have examined whether males and females differ in their sensitivity to alcohol. Mills and Bisgrove[17] found no gender differences in a divided-attention task after a low dose of alcohol, but greater impairment in females at a higher dose. However, in this study women achieved higher BACs, and reported less alcohol consumption compared to men. Burns and Moskowitz[57] found no significant gender differences in a series of motor and cognitive impairment tasks. When controlling for BACs, Niaura et al.[11] found few gender differences in psychomotor and cognitive responses to alcohol. Other research that controlled for gender differences in BACs and drinking practices has found few gender differences in response to alcohol.[58] Overall, when controlling for BACs and drinking practices, gender differences in alcohol impairment have not been demonstrated.

Little research has examined differences in alcohol impairment that are related to age. Using groups with fairly equivalent drinking practices, Parker and Noble[59] found that older subjects (over 42 years old) had more deficits on abstracting and problem solving after alcohol consumption compared to younger subjects. Linnoila et al.[8] found a trend toward increased impairment in subjects who were 25 to 35 years old, compared to those 20 to 25 years old. Other studies have also found age-related increases in psychomotor impairment in humans[60,61] and in animals,[62] even when BACs and drinking practices were equivalent in older and younger groups. Although there appears to be an increase in sensitivity to alcohol's effects with advancing age, there are few studies, and most suffer from small sample sizes. Because age effects appear to occur even when BACs and drinking practices are controlled, some have speculated that age-related increases in alcohol impairment reflect the effects of aging on vulnerability of the central nervous system to alcohol's effects.[62]

1.3 TIME OF INGESTION

The effects of alcohol tend to vary dramatically over the time course of a drinking episode. An analysis of how the effects of alcohol change over time provides a clearer understanding of alcohol-related impairment. This chapter reviews differences in the effects of alcohol during the rising and falling limbs of the BAC curve, the phenomenon of acute tolerance, and post-drinking hangover.

1.3.1 Rising and Falling Blood Alcohol Concentrations

Researchers and clinicians have long noted that alcohol's effects are often biphasic during a drinking episode.[63] "Biphasic" refers to the fact that stimulant effects of alcohol tend to precede sedative alcohol effects during a drinking episode.[64,65] There are substantial individual differences in the magnitude of stimulant effects of alcohol and the BACs at which they occur.[66,67] Alcohol's stimulant effects are reflected in increased motor activity, talkativeness, and euphoric or positive mood at lower doses and during rising BACs.[67,68] Stimulant effects have been assessed in humans using a variety of psychophysiological, motor activity, and self-report measures,[65,66,69] and in animals using measures such as spontaneous motor activity.[63,70] Stimulant effects are present in some drinkers at BACs as low as 20 to 30 mg/dL, and may persist on the rising BAC limb well past 100 mg/dL.[63,65] Some current theories hold that stimulant effects reflect alcohol's reinforcing qualities, and that the magnitude of stimulant effects will predict future drinking and the development of alcohol dependence.[71,72]

Some research suggests that the rate of change of rising BACs helps determine the degree of alcohol effects. A faster rise of ascending BACs is associated with greater euphoria and intoxication,[71,73-75] as well as increased behavioral impairment.[11,76] It is interesting to speculate that the drinking patterns shown by many heavy drinkers and alcoholics may reflect an attempt to produce a rapid rise in BACs. These patterns include gulping drinks, drinking on an empty stomach, and using progressively fewer mixers to dilute distilled spirits.

Sedative effects of alcohol usually occur at higher BACs and on the descending limb of the BAC curve. Sedation has been measured in humans using EEG patterns and self-reports of anesthetic sensations and dysphoric mood,[64,74,77] and in animals with low motor activity and the onset of alcohol-induced sleep.[78,79] Robust sedative effects tend to first appear at peak BACs near 60 to 80 mg/dL in many drinkers, although in persons with higher tolerance sedative effects are not apparent until BACs are above 100 mg/dL.[49,64,65] Sedative effects of alcohol are negatively correlated with drinking practices, and lower levels of sedation after alcohol consumption may characterize persons at increased risk for the future development of alcoholism.[79-81]

Research has clearly demonstrated that alcohol-related impairment is greater on the ascending compared to the descending limb of the BAC curve. This finding appears consistently across different doses and impairment tests. The most straightforward explanation for this effect is acute tolerance.

1.3.2 Acute Tolerance

There are many different types of tolerance identified by researchers.[50] Metabolic tolerance refers to an acquired increase in the rate of alcohol metabolism. Functional tolerance can be defined as an acquired decrease in an effect of alcohol at a given BAC. There are several different types of functional tolerance. Chronic tolerance refers to an acquired decrease in an effect of alcohol across multiple exposures to the drug. This chapter focuses on acute tolerance, defined as a decrease over time in an effect of alcohol within a single exposure to alcohol, which occurs independently of changes in BAC.

Acute tolerance is one of the most robust effects that occur in laboratory alcohol administration research. In 1919, Mellanby[82] demonstrated that effects of alcohol were greater during the rising compared to the falling limb of the BAC curve, a phenomenon known as the "Mellanby effect." A number of laboratory studies in humans and animals have replicated the Mellanby effect using numerous measures, such as motor coordination, self-reported intoxication, sleep time, and body temperature.[12,83-85]

One early question raised about the Mellanby effect was whether it was the result of a methodological artifact in the measurement of alcohol concentration in blood. Venous BAC, which is sampled for alcohol measurement, is known to lag behind arterial BAC during the ascending limb of the BAC curve, before the distribution of alcohol throughout body water compartments is complete. It is arterial blood that is closest to brain levels of alcohol. Thus, some wondered whether the Mellanby effect was an artifact because it actually compared impairment at different concentrations of alcohol in the brain.

It has been established, however, that acute tolerance and the Mellanby effect occur beyond any differences between arterial and venous BAC. First, the Mellanby effect is robust when BACs are assessed via breath alcohol; breath measures are closer to arterial than to venous BACs during the ascending limb of the BAC curve. Second, researchers have demonstrated the presence of acute tolerance using numerous alternative methods. When BACs are at a steady state, acute tolerance has been demonstrated by decreases in the effects of alcohol that occur over time.[13,86] Furthermore, the rate of decrease in alcohol effects are significantly greater than the rate of decrease in descending BACs.[87–89] Another demonstration of acute tolerance comes from social drinkers who report themselves as feeling completely sober when descending BACs are substantial (e.g., 30 to 50 mg/dL).[87] Investigators have demonstrated that decreased effects over time within an exposure to alcohol are not due to practice or other repeated-measures effects.[90-92]

In some models, acute tolerance occurs in a linear fashion as a function of the passage of time, independent of alcohol concentration.[89] Others have proposed that acute tolerance is concentration dependent.[93] Even in this latter model, the passage of time is critical; alcohol concentration simply influences the rate of recovery over time. Martin and Moss[87] found that at both higher and lower doses of alcohol, the magnitude of the Mellanby effect was highly correlated with the amount of time between the ascending and descending limb measurements. Clearly, the amount of impairment and intoxication shown during a drinking episode is not only affected by the level of BAC, but also by the amount of time alcohol has been in a person's system.

Vogel-Sprott and colleagues have published a large body of research that demonstrates that the degree of acute tolerance development is influenced by rewards and punishments, i.e., the consequences of non-impaired and impaired performance.[94] These investigators have demonstrated that acute tolerance to the impairment produced by alcohol increases when non-impaired performance leads to financial reward or praise.[12,95] One way to interpret these findings is in terms of motivation to show non-impaired performance. When an intoxicated motorist is stopped for questioning and/or field sobriety tests, that person will be highly motivated to show non-impaired behavior. While such attempts at appearing sober will be unsuccessful when BACs are sufficiently high, it is likely that many motorists avoid the detection of intoxication at substantial BACs by being highly motivated. Unfortunately, the same motorists may again show substantially impaired performance after the immediate contingency of detection and arrest are removed. Most of the laboratory impairment studies reviewed here did not adequately control for or study the impact of high motivational levels on the obtained results. For this reason, it is likely that the magnitude of impairment observed in many laboratory studies is greater than would be obtained in a field setting.

1.3.3 Hangover

Hangover has received relatively little attention in the scientific literature, but it can certainly produce alcohol-related impairment. Hangover is an aversive state typically experienced the morning after a heavy drinking bout, which is characterized by dysphoric and irritable mood, headache, nausea, dizziness, and dehydration. Sufficiently heavy drinking produces self-reported hangover symptoms in most persons, but there appears to be large individual differences in the occurrence, severity, and time course of perceived hangover that are independent of drinking practices.[96]

Studies reported in the literature are contradictory concerning whether hangover is accompanied by behavioral impairment. Several early studies found no performance impairment when BACs were at or close to zero.[97,98] Myrsten and colleagues[99] tested a variety of behavioral impairment measures 12 hours after subjects consumed a dose of alcohol producing mean BACs near 120 mg/dL. Morning BACs in this study averaged 4 mg/dL. Most measures showed no impairment, but hand steadiness was detrimentally affected. Collins and Chiles[100] gave impairment tests before an evening drinking session, after alcohol consumption, and again after subjects had slept 4 to 5 hours. Subjects were affected on the performance tests during acute intoxication, and they reported significant levels of hangover during the morning session. Despite these methodological strengths, there were no clear-cut performance impairing effects of hangover in this study.

Some other research indicates that hangover is accompanied by impairment in behavioral and cognitive functioning. One experiment found that after high peak BACs of about 150 mg/dL, there was an average 20% decrement on a simulated driving task 3 hours after BACs had returned to zero.[101] Yesavage and Leirer[102] examined hangover effects in Navy aircraft pilots 14 hours after drinking enough alcohol to produce BACs of about 100 mg/dL, using a variety of flight simulator measures. Significant detrimental effects of hangover were found for 3 of 6 variance measures and 1 of 6 performance measures. Similarly, other studies have found impairment due to hangover on only some of the tests employed in the research.[103,104]

Some have speculated that the nausea and dizziness of hangover may be associated with PAN II, an eye movement nystagmus that occurs during falling BACs and after measurable alcohol has left the system.[35] PAN II reflects the sensitivity of semicircular canal receptors to gravity, which also produces feelings similar to motion sickness. More research is needed to address the role of vestibular functioning in hangover.

Inconsistencies in the hangover literature probably reflect the fact that behavioral impairment during hangover is influenced by numerous factors, including sleepiness, fatigue, mood, and motivation to behave in a non-impaired manner. Moreover, there are individual differences in the frequency and duration of hangover, even among persons with similar drinking practices. For many persons, the BACs required to produce subsequent hangover may be greater than those typically obtained in laboratory studies. Hangover effects are poorly characterized, and are an important topic for further research.

1.4 IMPAIRMENT TESTING

1.4.1 Ideal Characteristics of Impairment Tests

The strengths and weaknesses of impairment tests used in the laboratory and the field are best evaluated in contrast with their ideal properties. There are a number of concepts that can be used to describe the characteristics of impairment tests, including scaling of results, applicability to field settings, reliability, validity, sensitivity, and specificity.[105]

Impairment tests differ in the scaling of results, that is, the nature of the scores or outcomes of a test. Results may be binary (impaired or not), ordinal rankings (low, medium, or high impairment), or quantitative scores. The need for precision of results depends upon the testing application. When used primarily as a screening tool for other sobriety tests or a BAC assessment, binary scores or ordinal rankings may be adequate. In other instances, continuous scores are desirable because they inform about the level of impairment.

Applicability to field settings is important for any impairment test used in law enforcement. One requirement is that a test must have adequate measurement properties in a field setting. A test may have demonstrated reliability and validity in the laboratory, but these properties may or may not generalize to field settings. Field applications involve a loss of control over numerous variables that can influence testing. Reliability and validity properties in the field may be far different from the laboratory, in part because data from a known sober condition are not available for comparison.

There are several other important considerations in relation to field settings. Ideally, a test must be easily administered in a standard way by test administrators, and readily understood by test takers. The level of technical skill required for administering the test, collection of data, and interpretation of results should be acquired with a reasonably short duration of training. Any required instrumentation should not require extensive maintenance, and should not be easily subject to interference by test takers. Importantly, impairment tests must have credibility with law enforcement officers and the wider criminal justice system. Whereas sobriety tests are often used as a preliminary screen for reasonable cause in BAC testing, they nevertheless must be generally acceptable to prosecutors, judges, and juries.

1.4.1.1 Reliability

Reliability refers to the extent to which a test provides a result that is stable or repeatable. That is, a reliable test is one that will yield a similar result across multiple testings in the same person (test-retest reliability), or across multiple test administrators or raters examining the same person (inter-rater reliability). An ideal impairment test should be reliable across testings in a sober person, i.e., it would reveal a stable baseline for non-impaired performance. Furthermore, an ideal test should show reliability across testings in an intoxicated person. That is, multiple tests taken at the same level of impairment would show relatively little variation in the obtained scores. Reliability is a key feature of any impairment test. If obtained scores are not reliable, the results may be caused by factors other than impairment, such as variation in test administration or scoring.

1.4.1.2 Validity

Validity concerns the extent to which a test accurately measures what it was intended to measure. The validity of impairment tests refers to the extent to which these tests assess alcohol-related impairment, rather than other factors. Face validity refers to the extent to which law enforcement personnel and test takers believe that a test does measure alcohol impairment; many field sobriety tests have high levels of face validity. Concurrent validity refers to the extent to which an impairment test shows expected associations with other tests known to measure impairment, and with BACs. Construct validity refers to the adequacy and explanatory power of scientific concepts such as alcohol impairment. If alcohol-related impairment is highly variable across different behaviors, this will reduce the validity of any one test in measuring such a diffuse concept.

A test can be reliable but not valid. For example, a person's height can be measured in a highly reliable fashion, but the observed results would be an invalid index of alcohol impairment. In contrast, some level of reliability is needed in order for a test to show validity. If an impairment test has no reliability, it cannot be valid. The degree to which reliability is imperfect tends to place an upper limit on the degree of validity that can be shown by a test.[106]

1.4.1.3 Sensitivity

Sensitivity refers to the ability of a test to detect impairment, and can be defined as the proportion of impaired persons (as determined by some other established measure) who are classified as impaired by a test. Thus, insensitive tests can allow persons who are impaired to escape detection (a false-negative test result). Whereas signs of intoxication are evident from test results in almost all persons when BACs become sufficiently high, many measures do not detect impairment when BACs are below 100 mg/dL. Among some heavy drinkers, many tests will be insensitive to BACs well above 100 mg/dL. In some cases, there probably is little impairment to detect when a test does not reveal impairment. In other cases, however, impaired performance is likely present, but a test is not sensitive enough to detect it.

1.4.1.4 Specificity

The specificity of an impairment test refers to the extent to which the results reflect alcohol impairment and not other factors such as fatigue, stress, and individual differences in cognitive and motor skills. A highly specific test will not be much influenced by changes in parameters other than alcohol impairment. An example of high test specificity in biological measurement is seen for BACs, where existing instrumentation allows assessment of alcohol in blood and breath that is not affected by closely related chemical compounds such as acetate, acetaldehyde, or acetone. Tests with low levels of specificity will lead to a high proportion of false-positive test identifications. That is, results for a non-specific test will often suggest that a person is impaired when he actually is not impaired (as measured by BAC or other tests). Thus, low specificity in impairment testing can lead to an inefficient expenditure of law enforcement resources.

An important issue in evaluating the sensitivity and specificity of impairment tests is whether measures of sober performance are available. Many tests show large individual differences in sober performance.[13,94] In laboratory research tests, sensitivity and specificity are evaluated by comparing a subject's performance at different BACs with their test performance when sober, usually before drinking begins. However, sensitivity and specificity are more difficult to achieve in the field, when sober baseline performance data are not available. Therefore, tests that are known to have less variation among sober persons are preferable for field settings.

When developing cutoff scores on impairment tests, increased sensitivity almost always comes at the expense of decreased specificity, and vice versa. The "best" cutoff score for the definition of impairment depends upon the relative importance of sensitivity and specificity in a given application, as well as the estimated base rate of impairment in the population that will be tested.[106] The choice of an appropriate cutoff score for an impairment test must be based on an understanding of these factors.

1.4.2 Characteristics of Existing Field Sobriety Tests

This chapter focuses on three field sobriety tests (FSTs) that have been standardized by the National Highway Traffic Safety Administration,[107] and which are widely used in the U.S. and elsewhere. In the one-leg stand test, subjects must raise one foot at least 6 in. off the ground and stand on the other foot for 30 seconds, while keeping their arms at their sides. Performance is scored on a 4-point scale, using items such as showing significant sway, using arms for balance, hopping, and putting down the raised foot. In the walk-and-turn test, subjects must balance with feet heel-to-toe and listen to test instructions. Then, subjects must walk nine steps heel-to-toe on a straight line, turn 180 degrees, and walk nine additional steps heel-to-toe, all the time counting their steps, watching their feet, and keeping their hands at their sides. Performance is scored on an 8-point scale, using items such as starting before instructions are finished, stepping off the line, maintaining balance with arm movements, and taking an incorrect number of steps. The gaze nystagmus test assesses horizontal gaze nystagmus. The angle of onset of nystagmus is assessed for each eye. Performance is scored on a 6-point scale (3 possible points for each eye).

Some research has examined the properties of these three FSTs. In a laboratory study, Tharp et al.[42] used 297 drinking volunteers with BACs from 0 to 180 mg/dL who were tested by police officers trained in the use of FSTs. Inter-rater reliability correlations for the FSTs ranged from 0.60 to 0.80, indicating an adequate level of reliability across test administrators. Test-retest correlations, examining the correspondence of FST scores on two separate occasions at similar BACs, ranged from about 0.40 to 0.75, indicating adequate test-retest reliability. All of the FSTs correlated significantly with BACs. Using all three FST test scores, officers were able to classify 81% of persons in terms of whether their BACs were above or below 100 mg/dL. Similar results using these standard FSTs in a field setting were obtained by Anderson et al.[108] However, neither of these reports provided data on the specificity and sensitivity of individual FSTs.

Few studies have reported the characteristics of individual FSTs in field settings. One study found that HGN, specifically the angle of horizontal nystagmus onset, had high levels of sensitivity and specificity in predicting BACs above 100 mg/dL in an emergency room setting.[43] Perrine et al.[109] examined the reliability and validity of the National Highway Traffic Safety Administration FSTs in a field setting with 480 subjects, using police officers and other trained individuals as test administrators. Inter-rater reliability was adequate for all three FSTs. All of the FSTs were significantly correlated with BAC; however, the magnitude of these correlations was low in the case of the walk-and-turn and the one-leg stand tests. Perrine et al.[109] provided data on the sensitivity and specificity of each FST as a function of different levels of BAC. The data indicated that the horizontal gaze nystagmus test had excellent sensitivity and specificity characteristics. Only 3% of subjects with a zero BAC failed the horizontal gaze nystagmus test (i.e., specificity was high when referenced to a zero BAC). Sensitivity was 100% for those with BACs over 150 mg/dL, and was 81% for subjects with BACs ranging from 100 mg/dL to 149 mg/dL.

However, Perrine et al.[109] found that prediction of BAC was much worse using the walk-and-turn and the one-leg stand test. For the walk-and-turn test, specificity was low, in that about half of those with a zero BAC were classified as impaired. While sensitivity was fairly high (78%) at BACs above 150 mg/dL, this parameter fell below 50% for those with BACs between 80 mg/dL and 100 mg/dL. For the one-leg stand test, 30% of those with a zero BAC were classified as impaired, indicating only moderate specificity. Sensitivity was only 50% for those with BACs between 100 mg/dL and 150 mg/dL, but improved to 88% for those with BACs above 150 mg/dL.

The literature suggests that overall, FSTs such as the walk and turn, one leg stand, and horizontal gaze nystagmus test, have adequate properties for detecting impairment in field settings. These tests meet the requirements of applicability to the field, in that they can be administered in a standardized fashion, have face validity, and are understood by test takers. Training of test administrators can occur in a reasonable period of time. Results suggest that these tests can be administered reliably, and have validity in the sense that they do measure impairment due to alcohol. However, horizontal gaze nystagmus performs much better at predicting BAC than the walk-and-turn and the one-leg stand tests. While they are worthwhile, the latter two FSTs have significant limitations when used in field settings. More research is needed to determine if improved testing procedures or different cutoff scores can improve the performance of the walk-and-turn and the one-leg stand tests.

It is important to note that the properties of FSTs depend upon the threshold BACs they are supposed to detect. FSTs can be used to test impairment that occurs at BACs below a legal limit, but much of their utility depends upon their ability to determine whether a person has a BAC above that allowed by law. The FSTs described above were designed and tested in the context of the limits of 100 mg/dL that exist throughout most of the U.S. However, the limit is currently 80 mg/dL in California, and is 50 mg/dL or 20 mg/dL in many European countries. It is likely that the performance of FSTs decreases as the BAC limit decreases. More research is needed to determine whether FSTs have utility when used to detect lower BAC threshold limits.

1.5 CONCLUSIONS

While it is clear that alcohol impairs performance, the presence and degree of impairment depends upon a large number of individual, situational, and pharmacological factors. Moreover, impairment is not uniform across all types of behavioral and cognitive functioning. Simple behaviors, such as reaction time tasks performed in isolation, are generally insensitive to alcohol consumption. Well-practiced behaviors tend to be insensitive to alcohol except at very high doses.

Impairment is seen consistently in tasks requiring the simultaneous processing of multiple sources of information. The results from dual-performance and divided-attention tasks suggest that alcohol reduces the amount of limited-capacity attentional resources available for coordinating multiple tasks. These results provide an important view of how alcohol can produce accidents and injuries. For example, the intoxicated motorist may perform with only moderate impairment on a well-known route with little traffic. However, when a situation arises that requires the simultaneous processing of multiple sources of information, such as avoiding an unexpected obstacle in traffic, large performance deficits may occur.

Alcohol also produces deficits in activities that require fine motor control at high rates of speed. One of the most sensitive behavioral measures of impairment found in the literature is tracking performance, which requires rapid small adjustments in the muscles of the hands and eyes, and a high level of hand-eye coordination. Tracking performance is an important aspect of impairment, in part, because it is central to the lateral guidance of a motor vehicle. Another type of behavior sensitive to alcohol is speech, which requires fine motor control, timing, and coordination among the lips, tongue, and vocal cords at a high rate of speed.

Impairments in eye movements and balance after drinking primarily reflect alcohol's effects on the brain's vestibular system. Alcohol is a small water-soluble molecule that readily diffuses throughout the brain, and impairment is often described in terms of alcohol's general depressant effects on all neural functions. However, impairment in vestibular functioning provides an example of specificity in alcohol's effects. The mechanisms of vestibular impairment have been fairly well characterized, and relate to how alcohol affects the specific physiology of this system. Vestibular functioning is relatively sensitive to alcohol's effects and important for behavioral functioning, and therefore is a logical focus of impairment testing.

There are large individual differences in the impairment produced by alcohol consumption, even when controlling for level of BAC. Several sources of these individual differences have been identified. Perhaps the most important factor is drinking practices. Those who drink more often and in greater amounts tend to develop a greater degree of chronic tolerance to the impairing effects of alcohol, i.e., have an acquired decrease in the degree of impairment across multiple drinking sessions. However, by definition, heavy drinkers tend to consume more and will be more likely to have higher BACs when tested in forensic settings. Thus, heavier drinking practices are probably not predictive of less impairment in field settings because individual differences in BACs are not controlled. Impairment tends to be greater in older adults as compared to younger adults; this effect is likely a combination of increased neural vulnerability with aging and differences in drinking practices between young and old. When controlling for BACs and drinking practices, gender differences in impairment have not been demonstrated.

The impairment produced by alcohol depends upon the time course of a drinking episode. Alcohol's effects have been described as biphasic, with initial euphoria and stimulant effects during early rising BACs, followed by dysphoria and sedative effects later on. Numerous measures of impairment are greater on the ascending compared to the descending limb of the BAC curve. This change in impairment due to limb of the BAC curve most likely reflects the phenomenon of acute tolerance, in which alcohol effects decrease over time within a drinking episode. In the prediction of impairment, the amount of time elapsed since alcohol has been in the system can be as important a variable as BAC itself.

The time course of alcohol's effects does not always end when BACs fall to zero. Sufficient drinking can produce hangover in many persons. Hangover is accompanied by impairment in behavioral functioning in some studies. Other research, however, has not found consistent effects of hangover on performance. There appear to be large individual differences in the degree of hangover effects and their duration. Hangover effects are poorly characterized compared to other effects of alcohol, and are an important topic for further research. The demonstration of significant impairment related to hangover would suggest the need for a longer period of abstinence from alcohol use before job performance in some professions, similar to the rules often applied to airline pilots.

Field sobriety tests, such as the walk and turn, one-leg stand, and horizontal gaze nystagmus tests, can be administered in a standardized fashion and meet the requirements of applicability to the field. These tests have adequate levels of validity, in that they reliably assess functions known to reflect alcohol impairment, and correlate with BACs and other impairment measures. Levels of sensitivity and specificity appear to be fairly high for the horizontal gaze nystagmus test. On the other hand, levels of test sensitivity and specificity are adequate but somewhat low for the walk-and-turn and the one-leg stand. That is, many persons with positive BACs, including those over 100 mg/dL, will be classified as not impaired using these tests, and many persons with low or zero BACs will be classified as impaired. Furthermore, the validity of FSTs for detecting lower threshold BACs such as 50 mg/dL or 20 mg/dL remains to be established. The development of new field sobriety tests that increase the accurate assessment of impairment would be of great benefit in forensic applications.

REFERENCES

1. Carpenter, J. A., Effects of alcohol on some psychological processes, *Quarterly Journal of Studies on Alcohol*, 23, 274, 1980.
2. Levine, J. M., Kramer, J., Levine, E., Effects of alcohol on human performance, *Journal of Applied Psychology*, 60, 508, 1975.
3. Finnigan, F., Hammersley, R., The effects of alcohol on performance, in *Handbook of Human Performance*, Volume 2, Smith, A., Jones, D., Eds., Academic Press Ltd., Orlando, FL, 1992, p. 73.
4. Goldberg, L., Quantitative studies on alcohol tolerance in man. The influence of ethyl alcohol on sensory, motor, and psychological functions referred to blood alcohol in normal and habituated individuals, *Acta Physiol. Scand.*, 5 (Suppl. 16): 1, 1943.
5. Baylor, A. M., Layne, C. S., Mayfield, R. D., Osborne, L., Spirduso, W. W., Effects of ethanol on human fractionated response times, *Drug and Alcohol Dependence*, 23, 31, 1989.
6. Taberner, P. V., Sex differences in the effects of low doses of ethanol on human reaction time, *Psychopharmacology*, 70, 283, 1980.
7. Maylor, E. A., Rabbitt, P. M., James, G. H., Kerr, S. A., Effects of alcohol and extended practice on divided-attention performance, *Perception and Psychophysics*, 48, 445, 1990.
8. Linnoila, M., Erwin, C. W., Ramm, D., Cleveland, W. P., Effects of age and alcohol on psychomotor performance of men, *Journal of Studies on Alcohol*, 41, 488, 1980.
9. Fagan, D., Tiplady, B., Scott, D. B., Effects of ethanol on psychomotor performance, *British Journal of Anaesthesia*, 59, 961, 1987.
10. Golby, J., Use of factor analysis in the study of alcohol-induced strategy changes in skilled performance on a soccer test, *Perceptual and Motor Skills*, 68, 147, 1989.
11. Connors, G. J., Maisto, S. A., Effects of alcohol instructions and consumption rate on motor performance, *Journal of Studies on Alcohol*, 41, 509, 1980.
12. Beirness, D., Vogel-Sprott, M., Alcohol tolerance in social drinkers: operant and classical conditioning effects, *Psychopharmacology*, 84, 393, 1984.
13. Wilson, J., Erwin, G., McClearn, G., Effects of ethanol. II. Behavioral sensitivity and acute behavioral tolerance, *Alcoholism: Clinical and Experimental Research*, 8, 366, 1984.
14. Maylor, E. A., Rabbitt, P. M., Connolly, S. A., Rate of processing and judgment of response speed: comparing the effects of alcohol and practice, *Perception and Psychophysics*, 45, 431, 1989.
15. Moskowitz, H., Burns, M. M., Williams, A. F., Skilled performance at low blood alcohol levels, *Journal of Studies on Alcohol*, 46, 482, 1985.
16. Niaura, R. S., Nathan, P. E., Frankenstein, W., Shapiro, A. P., Brick, J., Gender differences in acute psychomotor, cognitive, and pharmacokinetic response to alcohol, *Addictive Behaviors*, 12, 345, 1987.
17. Mills, K., Bisgrove, E., Body sway and divided attention performance under the influence of alcohol: dose-response differences between males and females, *Alcoholism: Clinical and Experimental Research*, 7, 393, 1983.
18. Steele, C. M., Josephs, R. A., Drinking your troubles away. II. An attention-allocation model of alcohol's effects on stress, *Journal of Abnormal Psychology*, 97, 196, 1988.
19. Zador, P. L., Alcohol-related relative risk of fatal driver injuries in relation to driver age and sex, *Journal of Studies on Alcohol*, 52, 302, 1991.
20. Hindmarch, I., Bhatti, J. Z., Starmer, G. A., Mascord, D. J., Kerr, J. S., Sherwood, N., The effects of alcohol on the cognitive function of males and females and on skills relating to car driving, *Human Psychopharmacology*, 7, 105, 1992.
21. McMillen, D. L., Smith, S. M., Wells-Parker, E., The effect of alcohol, expectancy, and sensation seeking on driving risk taking, *Addictive Behaviors*, 14, 477, 1989.
22. Mongrain, S., Standing, L., Impairment of cognition, risk-taking, and self-perception by alcohol, *Perceptual and Motor Skills*, 69, 199, 1989.
23. Attwood, D. A., Williams, R. D., Madill, H. D., Effects of moderate blood alcohol concentrations on closed-course driving performance, *Journal of Studies on Alcohol*, 41, 623, 1980.
24. Huntley, M. S., Centybear, T. M., Alcohol, sleep deprivation and driving speed effects upon control use during driving, *Human Factors*, 16, 19, 1974.
25. Brewer, N., Sandow, B., Alcohol effects on driver performance under conditions of divided attention, *Ergonomics*, 23, 185, 1980.

26. Sobell, L. C., Sobell, M. B., Effects of alcohol on the speech of alcoholics, *Journal of Speech and Hearing Research*, 15, 861, 1972.
27. Sobell, L. C., Sobell, M. B., Coleman, R. F., Alcohol-induced dysfluency in nonalcoholics, *Folia Phoniatrica*, 34, 316, 1982.
28. Trojan, F., Kryspin-Exner, K., The decay of articulation under the influence of alcohol and paraldehyde, *Folia Phoniatrica*, 20, 217, 1968.
29. Lester, L., Skousen, R., The phonology of drunkenness, in *Papers from the Parasession on Natural Phonology*, Bruck, A., Fox, R. A., LaGay, M. W., Eds., Chicago Linguistic Society, Chicago, 1974, Chapter 8.
30. Pisoni, D. B., Martin, C. S., Effects of alcohol on the acoustic-phonetic properties of speech: perceptual and acoustic analyses, *Alcoholism: Clinical and Experimental Research*, 13, 577, 1989.
31. Johnson, K., Pisoni, D. B., Bermacki, R. H., Do voice recordings reveal whether a person is intoxicated? A case study, *Phonetica*, 47, 215, 1990.
32. Iurato, S., *Submicroscopic Structure of the Inner Ear*, Pergamon Press, London, 1967, p. 216.
33. Money, K. E., Johnson, W. H., Corlett, B. M., Role of semicircular canals in positional alcohol nystagmus, *American Journal of Physiology*, 208, 1065, 1965.
34. Nito, Y., Johnson, W. H., Money, K. E., The non-auditory labyrinth and positional alcohol nystagmus, *Acta Otolaryngology*, 58, 65, 1964.
35. Money, K. E., Myles, W. S., Hoffert, B. M., The mechanism of positional alcohol nystagmus, *Canadian Journal of Otolaryngology*, 3, 302, 1974.
36. Aschan, G., Gergstedt, M., Positional alcoholic nystagmus (PAN) in man following repeated alcohol doses, *Acta Otolaryngology*, Suppl. 330, 15, 1975.
37. Fregly, A. R., Bergstedt, M., Graybiel, A., Relationships between blood alcohol, positional alcohol nystagmus, and postural equilibrium, *Quarterly Journal of Studies on Alcohol*, 28, 11, 1967.
38. Aschan, G., Different types of alcohol nystagmus, *Acta Otolarnygology*, Suppl. 140, 69, 1958.
39. Behrens, M. M., Nystagmus, *Journal of Opthalmological Clinics*, 18, 57, 1978.
40. Lehti, H., The effect of blood alcohol concentration on the onset of gaze nystagmus, *Blutalkohol*, 13, 411, 1976.
41. Barnes, G. R., Crombie, J. W., Edge, A., The effects of ethanol on visual-vestibular interaction during active and passive head movements, *Aviation, Space, and Environmental Medicine*, July 1985, p. 695.
42. Tharp, V. K., Burns, M., Moskowitz, H., *Development and field test of psychophysical Tests for DWI Arrest: Final Report*, technical report DOT-HS-805-864, National Highway Traffic Safety Administration, Washington, D.C., 1981.
43. Goding, G. S., Dobie, R. A., Gaze nystagmus and blood alcohol, *Laryngoscope*, 96, 713, 1986.
44. Baloh, R. W., Sharma, S., Moskowitz, H., Griffith, R., Effect of alcohol and marijuana on eye movements, *Aviation, Space, and Environmental Medicine*, January 1979, p. 18.
45. Gentles, W., Llewellyn-Thomas, E., Effect of benzodiazepines upon saccadic eye movements in man, *Clinical Pharmacology and Theraputics*, 12, 563, 1971.
46. Niaura, R. S., Wilson, G. T., Westrick, E., Self-awareness, alcohol consumption, and reduced cardiovascular reactivity, *Psychosomatic Medicine*, 50, 360, 1988.
47. Lipscomb, T. R., Nathan, P. E., Wilson, G. T., Abrams, D. B., Effects of tolerance on the anxiety-reducing functions of alcohol, *Archives of General Psychiatry*, 37, 577, 1980.
48. Lipscomb, T. R., Nathan, P. E., Effect of family history of alcoholism, drinking pattern, and tolerance on blood alcohol level discrimination, *Archives of General Psychiatry*, 37, 576, 1980.
49. O'Malley, S. S., Maisto, S. A., Factors affecting the perception of intoxication: dose, tolerance, and setting, *Addictive Behaviors*, 9, 111, 1984.
50. Goldstein, D. B., *Pharmacology of Alcohol*, Oxford University Press, New York, 1983.
51. Frezza, M., DiPadova, C., Pozzato, G., Terpin, M., Baraona, E., Lieber, C., High blood alcohol levels in women: the role of decreased gastric alcohol dehydrogenase activity and first-pass metabolism, *New England Journal of Medicine*, 322, 95, 1990.
52. Sutker, P. B., Goist, K., King, A., Acute alcohol intoxication in women: relationship to dose and menstrual cycle phase, *Alcoholism: Clinical and Experimental Research*, 11, 74, 1987.
53. Brick, J., Nathan, P. E., Shapiro, A. P., Westrick, E., Frankenstein, W., The effect of menstrual cycle on blood alcohol levels and behavior, *Journal of Studies on Alcohol*, 47, 472, 1986.

54. Sutker, P. B., Goist, K., Allain, A. N., Bugg, F., Acute alcohol intoxication: sex comparisons on pharmacokinetic and mood measures, *Alcoholism: Clinical and Experimental Research*, 11, 507, 1987.
55. Cole-Harding, S., Wilson, J., Ethanol metabolism in men and women, *Journal of Studies on Alcohol*, 48, 380, 1987.
56. Jones, B. M., Jones, M. K., Alcohol effects in women during the menstrual cycle, *Annals of the New York Academy of Sciences*, 273, 576, 1976.
57. Burns, M., Moskowitz, H., Gender-related differences in impairment of performance by alcohol, in *Currents in Alcoholism, Volume 3: Biological, Biochemical and Clinical Studies,* Sexias, F., Ed., Grune & Stratton, New York, 1978, p. 479.
58. Sutker, P. B., Allain, A. N., Brantley, P. S., Randall, C. L., Acute alcohol intoxication, negative affect, and autonomic arousal in women and men, *Addictive Behaviors*, 7, 17, 1982.
59. Parker, E. S., Noble, E. P., Alcohol and the aging process in social drinkers, *Journal of Studies on Alcohol*, 41, 170, 1980.
60. Jones, M. K., Jones, B. M., The relationship of age and drinking habits to the effects of alcohol on memory in women, *Journal of Studies on Alcohol*, 41, 179, 1980.
61. Vogel-Sprott, M., Barrett, P., Age, drinking habits, and the effects of alcohol, *Journal of Studies on Alcohol*, 45, 517, 1984.
62. York, J. L., Increased responsiveness to ethanol with advancing age in rats, *Pharmacology, Biochemistry, and Behavior*, 19, 687, 1983.
63. Pohorecky, L. A., Biphasic action of ethanol, *Biobehavioral Reviews*, 1, 231, 1977.
64. Martin, C. S., Earleywine, M., Musty, R. E., Perrine, M. W., Swift, R. M., Development and validation of the biphasic alcohol effects scale, *Alcoholism: Clinical and Experimental Research*, 17, 140, 1993.
65. Tucker, J., Vuchinich, R., Sobell, M., Alcohol's effects on human emotions: a review of the stimulation/depression hypothesis, *International Journal of the Addictions*, 17, 155, 1982.
66. deWit, H., Uhlenguth, E., Pierri, J., Johanson, C., Individual differences in behavioral and subjective responses to alcohol, *Alcoholism: Clinical and Experimental Research*, 11, 52, 1987.
67. Nagoshi, C., Wilson, J., One-month repeatability of emotional responses to alcohol, *Alcoholism: Clinical and Experimental Research*, 12, 691, 1988.
68. Freed, E., Alcohol and mood: an updated review, *International Journal of the Addictions*, 13, 173, 1978.
69. Newlin, D., Thomson, J., Chronic tolerance and sensitization to alcohol in sons of alcoholics, *Alcoholism: Clinical and Experimental Research*, 15, 399, 1991.
70. Waller, M., Murphy, J., McBride, W., Effect of low dose ethanol on spontaneous motor activity in alcohol-preferring and non-preferring lines of rats, *Pharmacology, Biochemistry, and Behavior*, 24, 617, 1986.
71. Stewart, J., deWit, H., Eikelboom, R., Role of unconditioned and conditioned drug effects in the self-administration of opiates and stimulants, *Psychological Review*, 91, 251, 1984.
72. Wise, R., Bozarth, M., A psychomotor stimulant theory of addiction, *Psychological Review*, 94, 469, 1987.
73. Connors, G. J., Maisto, S. A., Effects of alcohol instructions and consumption rate on affect and physiological sensations, *Psychopharmacology*, 62, 261, 1979.
74. Lukas, S., Mendelson, J., Benedikt, R., Instrumental analysis of ethanol-induced intoxication in human males, *Psychopharmacology*, 89, 89, 1986.
75. Martin, C. S., Earleywine, M., Ascending and descending rates of change of blood alcohol concentrations and subjective intoxication ratings, *Journal of Substance Abuse*, 2, 345, 1990.
76. Moskowitz, H., Burns M., Effects of rate of drinking on human performance, *Journal of Studies on Alcohol*, 37, 598, 1976.
77. Wilson, J., Nagoshi, C., Adult children of alcoholics: cognitive and psychomotor characteristics, *British Journal of Addiction*, 83, 809, 1988.
78. Engel, J., Liljequist, S., The involvement of different central neurotransmitters in mediating stimulatory and sedative effects of ethanol, in *Stress and Alcohol Use*, Pohorecky, L. A., Brick, J., Eds., Elsevier Biomedical, New York, 1983, p. 153.
79. Tabakoff, B., Hoffman, P., Tolerance and the etiology of alcoholism: hypothesis and mechanism, *Alcoholism: Clinical and Experimental Research*, 12, 184, 1988.

80. Gabrielli, W., Nagoshi, C., Rhea, S., Wilson, J., Anticipated and subjective sensitivities to alcohol, *Journal of Studies on Alcohol*, 52, 205, 1991.
81. Schuckit, M., Subjective responses to alcohol in sons of alcoholics and control subjects, *Archives of General Psychiatry*, 41, 879, 1984.
82. Mellanby, E., *Alcohol: Its Absorption and Disappearance from the Blood under Different Conditions*, Great Britain Medical Research Council, Her Majesty's Statistical Office, London, 1919.
83. Gilliam, D., Alcohol absorption rate affects hypothermic response in mice: evidence for acute tolerance, *Alcohol*, 6, 357, 1989.
84. Martin, C. S., Rose, R. J., Obremski, K. M., Estimation of blood alcohol concentrations in young male drinkers, *Alcoholism: Clinical and Experimental Research*, 15, 494, 1990.
85. Waller, M., McBride, W., Lumeng, L., Li, T. K., Initial sensitivity and acute tolerance to ethanol in the P and NP lines of rats, *Pharmacology, Biochemistry, and Behavior*, 19, 683, 1983.
86. Kaplan, H., Sellers, E., Hamilton, C., Is there acute tolerance to alcohol at a steady state? *Journal of Studies on Alcohol*, 46, 253, 1985.
87. Martin, C. S., Moss, H. B., Measurement of acute tolerance to alcohol in human subjects, *Alcoholism: Clinical and Experimental Research*, 17, 211, 1993.
88. Nagoshi, C., Wilson, J., Long-term repeatability of alcohol metabolism, sensitivity, and acute tolerance, *Journal of Studies on Alcohol*, 50, 162, 1989.
89. Radlow, R., Hurst, P., Temporal relations between blood alcohol concentration and alcohol effect: an experiment with human subjects. *Psychopharmacology*, 85, 260, 1985.
90. Benton, R., Banks, W., Vogler, R., Carryover of tolerance to ethanol in moderate drinkers, *Journal of Studies on Alcohol*, 42, 1137, 1982.
91. Hurst, P., Bagley, S., Acute adaptation to the effects of alcohol, *Journal of Studies on Alcohol*, 33, 358, 1972.
92. LeBlanc, A., Kalant, H., Gibbons, R., Acute tolerance to ethanol in the rat, *Psychopharmacolgia*, 41, 43, 1975.
93. Kalant, H., LeBlanc, A., Gibbons, R., Tolerance to, and dependence on, some non-opiate psychotropic drugs, *Pharmacology Review*, 23, 135, 1971.
94. Vogel-Sprott, M., *Alcohol tolerance and social drinking: learning the consequences*, Guilford Press, New York, 1992.
95. Vogel-Sprott, M., Kartchner, W., McConnell, D., Consequences of behavior influence the effect of alcohol, *Journal of Substance Abuse*, 1, 369, 1989.
96. Newlin, D. B., Pretorious, M., Sons of alcoholics report greater hangover symptoms than sons of non-alcoholics: a pilot study, *Alcoholism: Clinical and Experimental Research*, 14, 713, 1990.
97. Collins, W. E., Schroeder, D. J., Gilson, R. D., Guedry, F. E., Effects of alcohol ingestion on tracking performance during angular acceleration, *Journal of Applied Psychology*, 55, 559, 1971.
98. Eckman, G., Frankenhaeuser, M., Goldberg, L., Hagdahl, R., Myrsten, A. L., Subjective and objective effects of alcohol as functions of dosage and time, *Psychopharmacologia*, 6, 399, 1964.
99. Kelly, M., Myrsten, A. L., Neri, A., Rydberg, U., Effects and after-effects of alcohol on psychological and physiological functions in man — a controlled study, *Blutalkohol*, 7, 422, 1970.
100. Collins, W. E., Chiles, W. D., Laboratory performance during acute alcohol intoxication and hangover, *Human Factors*, 22, 445, 1980.
101. Laurell, H., Tornros, J., Franck, D. H., If you drink, don't drive: the motto now applies to hangovers as well, *Journal of the American Medical Association Medical News*, October 7, 1983, p. 1657.
102. Yesavage, J. A., Leirer, V. O., Hangover effects on aircraft pilots 14 hours after alcohol ingestion: a preliminary report, *American Journal of Psychiatry*, 143, 1546, 1986.
103. Karvinen, E., Miettinen, A., Ahlman, K., Physical performance during hangover, *Quarterly Journal of Studies on Alcohol*, 23, 208, 1962.
104. Takala, M., Siro, E., Tiovainen, Y., Intellectual functions and dexterity during hangover, *Quarterly Journal of Studies on Alcohol*, 19, 1, 1958.
105. Allen, J. P., Litten, R. Z., Anton, R., Measures of alcohol consumption in perspective, in *Measuring Alcohol Consumption: Psychosocial and Biochemical Methods*, Litten, R. Z., Allen, J. P., Eds., Humana Press, Totowa, NJ, 1992, p. 205.
106. Meehl, P. E., Rosen, A., Antecedent probability and the efficiency of psychometric signs, patterns, or cutting scores, *Psychological Bulletin*, 52, 194, 1952.

107. National Highway Traffic Safety Administration, DWI Detection and Standardized Field Sobriety Testing: Administrators Guide, DOT-HS-178/RI/90, National Highway Traffic Safety Administration, Washington, D.C.
108. Anderson, T. E., Schweitz, R. M., Snyder, M. B., *Field Evaluation of a Behavioral Test Battery for DWI*, DOT-HS-806-475, National Highway Traffic Safety Administration, Washington, D.C.
109. Perrine, M. W., Foss, R. D., Meyers, A. R., Voas, R. B., Velez, C., Field sobriety tests: reliability and validity, in *Alcohol, Drugs and Traffic Safety-T92*, Utzelmann, H. D., Berghaus, G., Kroj, G., Eds., Verlag TUV Rheinland, Cologne, Germany, 1993.

CHAPTER 2

Update on Clinical and Forensic Analysis of Alcohol

Alan Wayne Jones, D.Sc.[1] and Derrick J. Pounder, M.D.[2]
[1] Department of Forensic Toxicology, University Hospital, Linköping, Sweden
[2] Department of Forensic Medicine, University of Dundee, Scotland, U.K.

CONTENTS

2.1 INTRODUCTION

Alcohol is the world's favorite recreational drug and moderate drinking has few untoward effects on a person's health and well-being.[1] Indeed, drinking small amounts of alcohol helps to relax people and relieve their inhibitions.[2] Moreover, scores of studies testify to the efficacy of small doses of alcohol, such as one to two glasses of red wine daily, as an effective prophylactic treatment for cardiovascular diseases such as ischemic stroke and heart failure.[1,3,4] In contrast, heavy drinking and drunkenness constitute major public health problems for both the individual and society.[5,6] Binge drinking is the cause of deviant behavior and is closely linked to family violence. Many of those who seek help from hospital casualty and emergency units are under the influence of alcohol.[7–10] High blood alcohol concentrations (BACs) are a common finding in all out-of-hospital deaths, especially in victims of suicide and drowning.[11–13] Accordingly, the determination of alcohol in body fluids is the most frequently requested service from forensic science and toxicology laboratories worldwide.[14–17]

The role of alcohol intoxication in traffic crashes and deaths on the roads is well recognized, which has led to the creation of punishable BAC limits for driving.[18–21] Measuring a person's blood- or breath-alcohol concentration (BrAC) furnishes compelling evidence for the prosecution case and, if above the legal limit, a guilty verdict is virtually guaranteed.[20,21] People who perform skilled tasks like operating machinery or other safety-sensitive work should avoid drinking alcohol for obvious reasons. Indeed, alcohol use in the workplace is regulated by statute in the U.S. (1991 Omnibus Transportation Employee Testing Act), and similar legislation can be expected in other countries.[22] The threshold BAC in connection with workplace testing is set at 20 mg/dL (0.02 g/210 L in breath), below which no action is taken. However, drinking on duty or having a BAC above 40 mg/dL (0.04 g/210 L of breath) is prohibited and the offending individual will be removed from participating in safety-sensitive work and risks, being dismissed.[22]

The punishment for driving under the influence of alcohol (DUI) includes a stiff fine, suspension of the driving license, and sometimes a period of imprisonment. Moreover, the validity of accident and insurance claims might be null and void if a person has been drinking and the BAC was above some threshold limit. The statutory alcohol limits for driving differ from country to country and this seems to depend more on political forces rather than traffic safety research and studies of crash risk as a function of BAC.[20,21] Table 2.1 lists the current legal alcohol limits in blood for driving in various parts of the world. These critical values are so-called per se concentration limits and additional proof that the person was under the influence of alcohol is unnecessary for a successful prosecution.[20]

The evolution of methods for determination of alcohol in body fluids has a long history and the first wet-chemical oxidation procedures were introduced more than 100 years ago.[17,23,24] Because of ease of collection and the larger volumes available, urine was the first biological specimen to be used for analysis of alcohol in clinical investigations. Finding a high concentration of alcohol

Table 2.1 Statutory BAC Limits for Driving in Different Parts of the World Expressed in Different Concentration Units

Country	g/100 mL	g/L (mg/mL)	mg/100 mL	mmol/L[a]
U.S. and Canada	0.08	0.80	80	17.3
Australia (most states)	0.05	0.50	50	10.9
U.K. and Ireland	0.08	0.80	80	17.3
Sweden and Norway[b]	0.02	0.20	20	4.3
Most EU countries	0.05	0.50	50	10.9

[a] Derived as [(mg/L)/46.07], where 46.07 is the molecular weight of ethanol to give mmol/L.

[b] The concentration unit mass/mass is used (mg/g or g/kg) so 0.02 mg/g = 0.21 mg/mL because the density of whole blood is 1.055 g/mL on average.

Table 2.2 Characteristic Features of Three Common Aliphatic Alcohols (Methanol, Ethanol, and Ethylene Glycol) Often Encountered in Forensic and Clinical Toxicology

Property	Methanol	Ethanol	Ethylene Glycol
CAS-number[a]	65-46-1	64-17-5	107-21-1
Molecular weight	32.04	46.07	62.07
Molecular formula	C_2H_4O	C_2H_6O	$C_2H_6O_2$
Chemical formula	CH_3OH	CH_3CH_2OH	$(CH_2OH)_2$
Structure	Primary aliphatic alcohol	Primary aliphatic alcohol	Dihydroxy aliphatic alcohol (diol)
Structural formula	H H—┼—OH H	H H H—┼—┼—OH H H	OH OH H—┼—┼—H H H
Common name	Wood alcohol	Beverage or grain alcohol	Antifreeze solvent
Boiling point, °C	64.7	78.5	197
Melting point, °C	–95.8	–114.1	–13
Density, at 20°C	0.791	0.789	1.11
Water solubility	Mixes completely	Mixes completely	Mixes completely
Main metabolites	Formaldehyde and formic acid	Acetaldehyde and acetic acid	Glycolaldehyde, glycolic, glyoxylic, and oxalic acid

[a] CAS, Chemical Abstract Service Registry Number.

in a sample of urine was a more objective test to verify clinical signs and symptoms of drunkenness. Moreover, many studies showed that the concentration of alcohol was highly correlated with subjective and objective measures of alcohol influence and performance decrement. The methods available for analysis of alcohol in biological specimens have become considerably refined and exhibit high sensitivity, specificity, accuracy, and precision.[14,15,25]

Fast and reliable analytical methods are needed in emergency situations, such as when a patient is admitted unconscious smelling of alcohol. One needs to distinguish gross intoxication from head trauma, which might require immediate surgery to remove intracranial blood clots.[26–28] Analytical methods used at hospital clinical laboratories need to differentiate ethanol intoxication from the impairment caused by drinking more dangerous alcohols like methanol and ethylene glycol.[29–31]

Table 2.2 gives basic chemical information about the alcohols most commonly encountered in clinical and forensic toxicology, namely, ethanol, methanol, and ethylene glycol. Depending on the concentration of methanol and ethylene glycol in blood, the emergency physician has to make a life saving decision to treat the poisoned patient with invasive therapy and antidotes.[32–34] This might entail administration of ethanol by intravenous infusion of a 10% solution to reach a BAC of 100 to 120 mg/dL and maintain this level for several hours. More recently the drug fomepizole (4-methyl pyrazole) has been introduced as an alternative treatment and is more suitable for use in young children or those with liver dysfunction.[32,35] Both ethanol and fomepizole function as competitive inhibitors of hepatic alcohol dehydrogenase (ADH), and help to prevent the conversion of methanol and ethylene glycol into their toxic metabolites, formaldehyde and formic acid and glycolic and oxalic acids, respectively.[35,36] The methanol and ethylene glycol remaining unmetabolized can be removed from the blood by hemodialysis and bicarbonate is usually administered to counteract acidosis caused by the acid metabolites of these more toxic alcohols.[33] The various treatment strategies currently available for dealing with methanol and ethylene glycol poisoning have been extensively reviewed.[29–36]

The diagnosis of drunkenness has broad social-medical ramifications and great care is needed when forensic practitioners and others are called upon to interpret results of analysis and draw conclusions about the degree of alcohol influence and the consequences for behavioral impairment. This chapter provides an update of clinical and forensic-medical aspects of alcohol analysis in body fluids, and research into the disposition and fate of alcohol in the body is also covered. In post-mortem toxicology, the choice of specimens, the preservation and storage, and particularly the interpretation of the results require special considerations. The role of alcohol in post-mortem toxicology with main focus on interpreting the analytical findings is covered in more detail in Chapter 3.

2.2 SPECIMENS FOR CLINICAL AND FORENSIC ANALYSIS OF ALCOHOL

2.2.1 Concentration Units

Unfortunately, there is no generally accepted way of reporting the results of alcohol analysis in body fluids and it is seemingly impossible to reach a consensus among scientists, scientific journals, and forensic practitioners on this issue.[37,38] Most clinical chemistry laboratories use the International system of units (SI system) established by broad international agreement. According to this system, the standard for the unit of mass is the kilogram, the unit of volume is the liter, and the amount of substance is the mole. The concentrations of endogenous or exogenous substances in serum or plasma are therefore reported as mmol/L or μmol/L. By contrast, forensic science and toxicology laboratories report their analytical results in mass per unit volume units (mg/dL, g/L, g/dL, or mg/mL) or in mass per unit mass units (g/kg or mg/g) when aliquots are weighed. The mass/mass unit is numerically less than the mass/volume unit by about 5.5% owing to the specific weight of whole blood, which is 1.055 g/mL on the average (1 mL whole blood weighs 1.055 g).[39] This means that a BAC of 100 mg/dL is the same as ~95 mg/100 g or 21.7 mmol/L.

2.2.2 Water Content of Serum and Whole Blood

The blood specimens received by forensic science and toxicology laboratories for analysis might be hemolyzed and are sometimes clotted. In contrast, the specimens analyzed at clinical laboratories usually consist of plasma or serum, that is, with the cellular elements, mainly red cells (erythrocytes), removed.[40] Because the water content of serum and plasma is higher than that of whole blood, the concentration of alcohol is also higher after removal of the red cells. Water content is a key factor controlling the distribution of alcohol in body fluids and this was recently the subject of a multicenter study in Germany.[41] The results reported by the three participating laboratories are compared in Table 2.3 and the values are given in mass/mass units, namely, g water per 100 g specimen.

Because this work was done at three different laboratories, each using highly reliably methods based on desiccation and gravimetric analysis, there is high confidence in the results reported.[42] To convert the results into mass/volume units, the values in Table 2.3 need to be multiplied by 1.055 (average density of whole blood is 1.055 g/mL).[39] The average serum/blood distribution ratio of water in this study was 1.157:1 and the standard deviation was 0.0163 ($N = 833$), with minimum and maximum values of 1.08 and 1.25. These results can be considered representative of people who drink and drive in Germany. Dividing the concentration of ethanol in serum by 1.16:1 gives the concentration expected in whole blood. Because plasma and serum contain the same amount of water, one expects the plasma/whole blood ratio of alcohol to be the same as the serum/whole blood ratio, as was shown empirically.[43] The results of alcohol analysis done at clinical chemistry laboratories should not be cited in drunken driving trials or other legal proceedings without an appropriate correction being made for the water content of the specimens or seeking expert help with interpretation of the results.[44,45]

The mean and standard deviation of the serum/whole blood distribution ratios of water from the German study[41] can be used to derive a 95% range of expected values in the relevant population as $1.16 \pm (1.96 \times 0.0163)$. Accordingly, 2.5% of individuals will have a serum/whole blood ratio above 1.19 and 2.5% will have a ratio below 1.12. Depending on requirements, more conservative

Table 2.3 Mean, SD, CV%, and Range of Water Content of Serum (A) and Whole Blood (B) and the Serum/Blood Ratio of Water Contents (C) Based on Measurements Made at Three Different Laboratories in Germany

Laboratory	N	Mean g%	SD g%	CV%	Range g%
		(A) Serum			
Kiel	230	90.49	0.86	0.95	87.2–93.3
Köln	503	90.71	0.54	0.59	88.6–92.2
Münster	100	90.75	0.61	0.67	89.1–92.8
Combined	833	90.66	0.66	0.72	87.2–93.3
		(B) Whole blood			
Kiel	230	78.35	1.44	1.83	74.8–82.9
Köln	503	78.41	1.11	1.41	75.7–83.0
Münster	100	78.14	1.28	1.63	74.9–83.3
Combined	833	78.36	1.23	1.57	74.8–83.3
		(C) Serum/Blood Ratio			
Kiel	230	1.156	0.0184	1.59	1.11–1.21
Köln	503	1.157	0.0145	1.25	1.10–1.20
Münster	100	1.162	0.0187	1.61	1.08–1.20
Combined	833	1.157	0.0163	1.40	1.08–1.21

values can be obtained by using the minimum and maximum values given in Table 2.3, namely, 1.08 and 1.21.[41]

2.2.3 Blood Hematocrit and Hemoglobin

Hematocrit is defined as the percentage in a volume of whole blood represented by the red blood cells (erythrocytes) and is determined by centrifugation. The hematocrit is sometimes referred to as the packed red cell volume, and a common reference interval for healthy men is 39 to 49% compared with 35 to 45% for healthy women with wider ranges in young and elderly individuals.[40] Because the water content of plasma is greater than that of whole blood and erythrocytes, one can expect that a blood specimen with high hematocrit (e.g., donated by a male subject) will contain less water than a blood specimen from a female with low hematocrit.[46] Extreme values in hematocrit and hemoglobin are likely to be found in conditions such as anemia (low values) or polycythemia (high values).[40,46]

Figure 2.1 (left graph) shows a strong positive correlation ($r = 0.96$) between blood hematocrit and hemoglobin content of whole blood specimens. The right plot shows a strong negative correlation ($r = -0.94$) between water content of whole blood and the hemoglobin content. These interrelationships will have a bearing on the plasma/whole blood and serum/whole-blood distribution ratios of alcohol in any individual case.

2.2.4 Alcohol Concentrations in Plasma and Whole Blood

Figure 2.2 shows mean concentration-time profiles of ethanol in plasma and whole blood after nine healthy men fasted overnight before drinking a bolus dose of ethanol (0.3 g/kg) diluted with orange juice.[47,48] The plasma curves run systematically above the whole-blood curves as expected from the difference in water content of the specimens. In this study, the experimentally determined plasma/whole-blood ratios of alcohol ranged from 1.08 to 1.19.[47] In a study with drinking drivers,

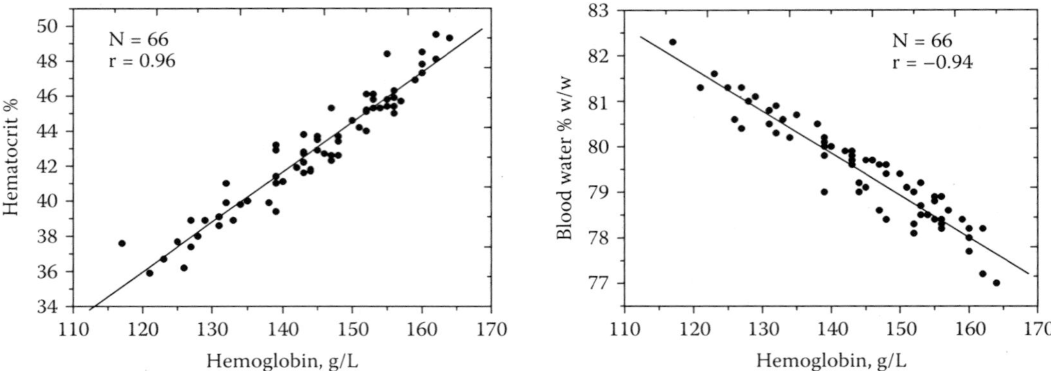

Figure 2.1 Strong positive correlation between hematocrit and hemoglobin content of venous blood samples (left plot) and strong negative correlation between water content and hemoglobin (right plot).

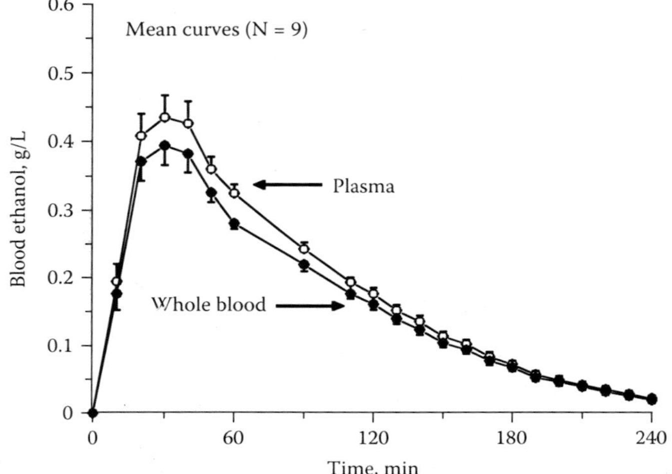

Figure 2.2 Mean concentration-time profiles of ethanol in specimens of whole blood and plasma from nine healthy men who drank a bolus dose of ethanol 0.30 g/kg mixed with orange juice in 15 min after an overnight fast.

a mean distribution ratio of ethanol between plasma and whole blood was reported to be 1.14:1 (standard deviation 0.041).[49]

Table 2.4 compares the alcohol concentrations in plasma or serum with values calculated by assuming a mean ratio of 1.16 and minimum and maximum values of 1.08 and 1.21 according to the above-referenced study from Germany (Table 2.3). The results are reported for ethanol concentrations ranging from 20 to 200 mg/100 mL.[41]

As discussed by Rainey,[44] whenever the concentration of alcohol in plasma or serum is used to estimate the concentration in whole blood for law enforcement purposes, it is advisable to consider the inherent variations in plasma/blood ratio of alcohol. He recommended that a conversion factor of 1.22:1 should be used, thus corresponding to mean + 2 SD from the various studies cited. This higher conversion factor seems more appropriate in forensic casework instead of using a mean value and will give a more conservative estimate of the person's BAC. In criminal law, a beyond-a-reasonable-doubt standard is necessary, whereas in civil litigation a preponderance of the evidence is sufficient to determine the outcome.

Table 2.4 Relationships between the Concentrations of Ethanol in Serum and the Expected Concentrations in Whole Blood Using Conversion Factors Based on Water Content of Serum and Whole Blood[41] (the 95% range and the minimum and maximum values are also calculated)

Serum Alcohol[a] mg/100 mL	Blood Alcohol Mean mg/100 mL	Blood Alcohol 95% Range	Blood Alcohol min; max
20	17.2	16.7–17.7	16.5; 18.5
50	43.1	41.9–44.3	41.3; 46.3
80	68.9	67.1–70.9	66.1; 74.1
100	86.2	83.9–88.6	82.6; 92.6
150	129.3	125.8–132.9	123.9; 138.9
200	172.4	167.7–177.3	165.3; 185.1

[a] Mean serum/blood alcohol ratio 1.16:1, standard deviation 0.0163, and minimum and maximum values 1.08 and 1.21 (see Table 2.3).

2.2.5 Allowing for Analytical Uncertainty

When blood samples are submitted for forensic alcohol analysis, it is a standard practice to make the determinations in replicate, always duplicate, and sometimes in triplicate.[50–52] Finding a close agreement between the two measurements (high precision) gives added assurance that no mishaps have occurred during sample handling and analysis. Besides reporting the average result, an allowance is necessary to compensate for uncertainty (random and systematic error) in the analytical method used.[25,51,52] This is easily done by making a deduction from the mean result to give a lower 95, 99, or 99.9% confidence interval depending on requirements.

In connection with prosecution for DUI in Sweden, besides the mean alcohol concentration, it is standard practice to report the lowest concentration of alcohol in the specimen analyzed with a probability of 99.9%.[25] The suspect's true BAC is therefore not less than this value with confidence of 99.9% or, in other words, there is a risk of 1:1000 of being unfair to the accused. The amount deducted from the mean of a duplicate determination is given by $[3.09 \times (SD/2^{1/2})]$, where SD is the standard deviation for a single determination using an analytical system demonstrated to be under statistical control.[25,51] The factor 3.09 is obtained from statistical tables of the normal distribution. Although the mean result of analysis is the best estimate of a person's BAC, the deduction compensates for factors beyond the suspect's control and gives the benefit of the doubt in those instances when the true BAC is close to the legal limit for driving. In such borderline cases, the risk of reporting a result, which in reality was just below the legal limit, as being above the limit is only 1 in 1000. The use of a deduction has practical significance only for individuals with a BAC at or near the critical legal limit for driving.

This use of a deduction is not customary in clinical chemistry laboratories, where single analyses are usually made.[53,54] Instead, most clinical chemistry laboratories compare their results with established reference ranges derived from previous experiences with specimens from healthy individuals depending on age and gender.[53] The result for the patient sample is compared with a so-called "normal range" in a comparable population of healthy individuals.[54] The imprecision of the analytical methods are monitored from the expected coefficient of variation (CV%) derived from analysis of calibration standards or spiked biological specimens analyzed along with the unknowns.[40,53]

2.3 MEASURING ALCOHOL IN BODY FLUIDS

2.3.1 Chemical Oxidation Methods

The first quantitative method of blood-alcohol analysis to gain general international acceptance in forensic science and toxicology was published in 1922[55] and became known as the Widmark micromethod (developed by Erik M.P. Widmark of Sweden). A specimen of capillary blood (100 mg) was sufficient for making a single determination and this could be obtained by pricking a fingertip or earlobe. Ethanol was determined by wet-chemistry oxidation with a mixture of potassium dichromate and sulfuric acid in excess. The amount of oxidizing agent remaining after the reaction was determined by iodometric titration.[55] Specially designed diffusion flasks permitted extraction of ethanol from the biological matrix by heating in a water bath for a few hours at 50°C. Ethanol and other volatiles were removed from the biological matrix by vaporization and oxidized within the flask and after cooling and addition of potassium iodide the final titrimetric analysis was done in the same flask. However, Widmark's method was not totally specific for determination of blood ethanol because other volatiles, if these were present, such as acetone, methanol, or ether, were also oxidized by the reagents and gave false high ethanol readings. The likelihood of potential interfering substances (e.g., acetone) could be tested by making a qualitative analysis of urine by adding various chemical reagents and looking for any characteristic color change such as that seen with high levels of ketones.[56]

By the 1950s the chemical oxidation methods were improved by monitoring the end point by photometry rather than by volumetric titration.[17,56] Today, analytical procedures based on wet-chemistry oxidation reactions are virtually obsolete in clinical and forensic laboratories for determination of alcohol in body fluids.[14] The history, development, and application of chemical oxidation methods of alcohol analysis have been well covered in several review articles.[17,57–59]

2.3.2 Enzymatic Methods

Shortly after the enzyme alcohol dehydrogenase (ADH) was purified from horse liver and yeast in the early 1950s the way was clear for developing biochemical methods to determine alcohol in body fluids.[60–63] These methods offered milder oxidation conditions and analytical selectivity was enhanced compared with wet-chemistry oxidation procedures. The ADH derived from human or animal liver proved less selective for oxidation of ethanol compared with the enzyme obtained from yeast.[62] Other aliphatic alcohols (methanol, isopropanol, and *n*-propanol) were oxidized by mammalian ADH but not acetone, which was the most problematic substance for the older wet-chemical methods.[64] By optimizing the test conditions in terms of pH, reaction time, and temperature, methanol was not oxidized by yeast ADH and this source of the enzyme is still used today for clinical and forensic alcohol analysis.[62,63]

A typical manual ADH method might entail precipitation of plasma proteins by adding perchloric acid and then adjusting the pH of the supernatant to 9.6 with semicarbazide buffer.[62] The purpose of the latter reagent, besides adjusting pH, was to trap the acetaldehyde produced during ethanol oxidation and thus drive the reaction to completion. The buffer is usually pre-mixed with the coenzyme NAD^+ before adding to the alcohol-containing supernatant, and finally the ADH enzyme is added to start the reaction. The amount of NAD^+ that becomes reduced to NADH is directly proportional to the concentration of ethanol in the original sample. After allowing the mixture to stand at room temperature for about 1 h to reach an end point, the NADH formed was determined by measuring absorption of ultraviolet (UV) light at a wavelength of 340 nm.

Later developments in ADH methods for analysis of alcohol tended to focus on separation of proteins from the blood by semipermeable membranes (dialysis) or by micro-distillation to obtain an aqueous ethanol solution for analysis.[65,66] With this modification and a Technicon® AutoAnalyzer device, several hundred blood samples could be analyzed daily.[65] Scores of publications have appeared describing diverse modifications and improvements to the original ADH method and dedicated "reagent kits" are commercially available. These kits were ideal for use at hospital laboratories and elsewhere where the throughput of samples was relatively low. Otherwise, most efforts were directed toward automating the sample preparation and dispensing reagents to increase sample throughput, and several batch analyzers or reaction rate analyzers appeared including a micro-centrifugal analyzer using fluorescence light scattering for quantitative analysis.[67,68]

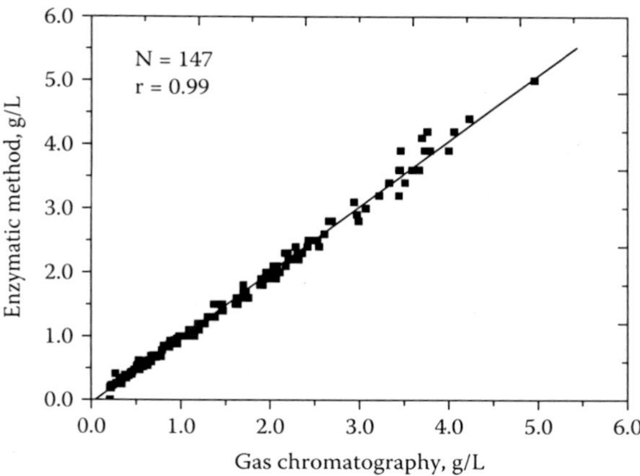

Figure 2.3 Conventional *x-y* scatter plot showing a high correlation between the concentrations of ethanol in urine determined by an automated enzymatic (ADH) method and by headspace gas chromatography (GC).

Enzymatic methods for determination of alcohol in body fluids are still used today in part owing to the widespread availability of multichannel analyzers for testing urine for drugs of abuse.[69] These procedures make use of a technique known as enzyme multiplied immunoassay (EMIT), whereby an enzyme-labeled antigen reacts with ethanol or another drug and the change in color after adding a substrate is measured by spectrophotometry.[69] The color intensity is related to the concentration of ethanol in the original specimen. Fluorescence polarization immunoassay (FPIA) and the spin-off technology radiative energy attenuation (REA) are other examples of analytical procedures developed to meet the increasing demand for drugs of abuse testing in urine and therapeutic drug monitoring.[70–73] In several comparative studies, excellent agreement was found for ethanol determined by REA and by gas chromatography in terms of accuracy and precision.[70,73] The principles and practice of various immunoassay systems suitable for clinical laboratory analysis were recently reviewed.[69]

Figure 2.3 shows a scatter plot comparing ethanol concentrations in urine determined by an automated ADH method and also by headspace gas chromatography. The correlation was excellent: $r = 0.99$. This kind of conventional x–y scatter plot is best redrawn and displayed as a Bland and Altman plot (Figure 2.4), whereby the difference in ethanol concentration by the two methods (GC–ADH) is plotted against the average concentration [(GC + ADH)/2]. The mean difference between the two methods indicates whether any bias exists between the two methods, which is a measure of accuracy. The SD of the differences gives the magnitude of scatter or variability of the individual differences and are referred to as limits of agreement between the two methods ($\pm 1.96 \times$ SD), shown as dotted horizontal lines on the plot. This Bland and Altman plot has gained considerable popularity for use in method comparison studies and makes it a lot easier to locate outlier values.[74] Three such outliers are circled and indicated on the plot; note that these occurred at very high BACs.

Despite many new developments in analytical technology for the analysis of alcohol in body fluids, particularly EMIT, FPIA, and REA methods, gas chromatography still dominates the instrument park at forensic toxicology laboratories owing to its superior selectivity. Indeed, some recent work has shown that elevated concentrations of serum lactate and lactate dehydrogenase might interfere with the analysis of alcohol by ADH methods.[75–77] This problem was traced to various side reactions whereby the coenzyme NAD^+ was reduced to NADH by endogenous substances and this could not be distinguished from NADH produced during the oxidation of ethanol. This resulted in undesirable false-positive results when plasma specimens from alcohol-free patients were analyzed.[76,77]

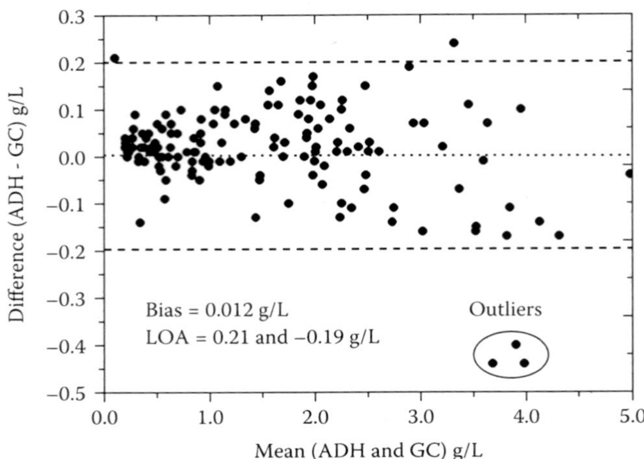

Figure 2.4 Bland and Altman plot of differences in results between the ADH and GC methods of analysis and the mean result of analysis by the two methods (data from Figure 2.3).

2.3.3 Gas Chromatographic Methods

In the early 1960s, physical-chemical methods began to be used for analysis of alcohol in body fluids such as infrared spectrometry, electrochemical oxidation, and gas-liquid chromatography (GLC).[78–83] Since then, GLC has become the method of choice for analysis of biological liquids in clinical and forensic laboratories. For determination of ethanol in breath, electrochemistry and infrared methods are widely used.[17] The first GLC methods required that the ethanol was extracted from blood by use of a solvent (e.g., *n*-propyl acetate) or by distillation, which was cumbersome and time-consuming.[78,79] Later developments meant that the blood was simply diluted (1:5 or 1:10) with an aqueous solution of an internal standard (*n*-propanol or *t*-butanol).[82,86] The five to ten times dilution with internal standard meant that matrix effects were eliminated and that aqueous alcohol standards could be used for calibration and standardization of the detector response.[82] The use of an internal standard also ensured that any unexpected variations in the GLC operating conditions during an analysis influenced the ethanol and the standard alike so the ratio of peak heights or peak areas (ethanol/standard) remained constant.[82]

The standard procedure entailed injecting 1 to 5 μL of the diluted blood into a heated chamber and any volatiles in the sample were mixed in a stream of helium or nitrogen, the carrier gas or mobile phase, which flowed through a glass or metal column with dimensions such as 2 m long by 0.3 mm inside diameter (i.d.). The column contained the liquid or stationary phase spread as a thin film over an inert solid support material, thus furnishing a large surface area. The volatile components of a mixture were distributed between the moving phase (carrier gas) and the liquid phase and depending on their physicochemical properties such as boiling point, functional groups present, and the relative solubility in the liquid phase, either partial or complete separation occurred during passage through the column. Polar stationary phases were an obvious choice for the analysis of alcohols, and polyethylene glycol with average molecular weights of 400, 600, 1500, etc. became widely available and were known as Carbowax phases.[83–86] Otherwise, porous polymer materials such as Poropak Q and S were useful as packing materials for the separating columns when low-molecular-weight, volatile-like alcohols were being analyzed.[86]

A quantitative analysis of ethanol required monitoring the effluent from the column as a function of time originally with a thermal conductivity (TC) detector,[78] but this was later replaced with a flame ionization detector (FID), which was more sensitive and gave only a very small response to water vapor in body fluids.[81,82] The ethanol concentration in blood was calculated by comparing

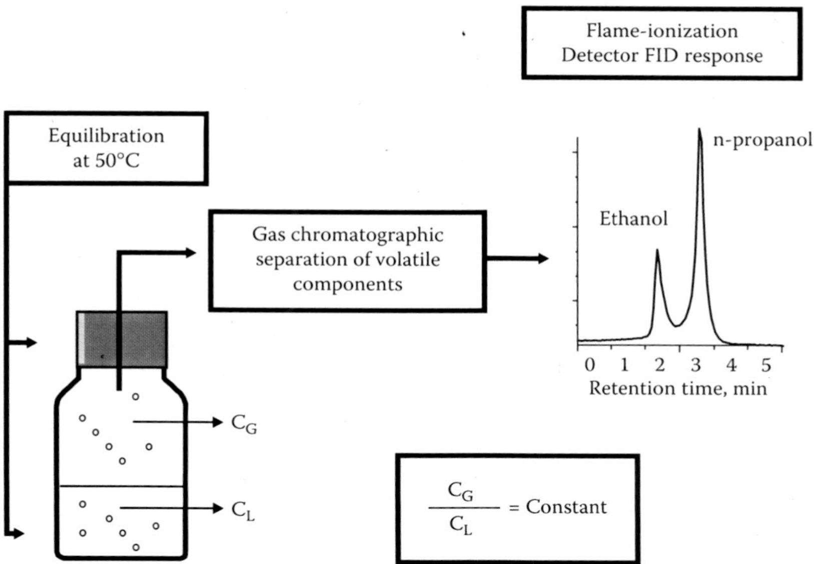

Figure 2.5 Schematic diagram of headspace gas chromatography. The blood or urine sample is first diluted 1:10 with an aqueous solution of an internal standard (*n*-propanol) directly into a glass vial made airtight with a crimped-on rubber membrane stopper. After reaching an equilibration at 50°C for 20 to 30 min, a sample of the vapor in equilibrium with the diluted specimen is removed either with a gas-tight syringe or by some automated sampling procedure and injected into the carrier gas (N_2) for transport through the chromatographic column to the detector. The resulting trace (chromatogram) shows a peak for ethanol followed by a peak for the internal standard (*n*-propanol). C_G is the ethanol concentration in the gas or air phase and C_L is the corresponding concentration in the liquid (blood or urine) phase at equilibrium.

the detector response (peak height or peak area) with results of analyzing known strength aqueous alcohol standards and making a calibration plot. Methodological details of many of the older GC methods of blood-alcohol analysis have been reviewed elsewhere.[84–86]

Figure 2.5 is a schematic diagram of the headspace gas chromatographic (GC-HS) analysis, which has become the method of choice in forensic science and toxicology laboratories for determination of ethanol and other volatiles in body fluids.[87–91] HS-GC requires that the blood samples and aqueous standards are first diluted (1:5 or 1:10) with an aqueous solution of an internal standard and the mixture kept airtight in a small glass vial with crimped-on rubber septum. The vials are then heated to 50 or 60°C for about 30 min to achieve equilibrium between concentrations in gas (C_G) and liquid (C_L) phases for all volatiles in the specimen. Care is needed not to heat the sample for too long at 60°C, otherwise some of the ethanol might be oxidized into acetaldehyde by a non-enzymatic oxidation reaction involving oxyhemoglobin.[92] This undesirable effect can be avoided by pretreating the blood specimen with sodium azide or sodium dithionite, chemicals that block this oxidation reaction.[89,92] However, it is simpler to work with a lower equilibrium temperature (40 or 50°C), which also prevents this oxidation reaction.[89]

The headspace vapor can be injected into the gas chromatograph manually with the aid of a gas-tight syringe or, as is more usual, an automated sampling procedure. Automation is preferred because this gives a much more reproducible injection and thus a higher analytical precision. Dedicated equipment for GC headspace analysis has been available since the early 1960s and the U.S. company Perkin-Elmer has dominated the market. Various versions of its headspace instruments have appeared over the years and in chronological order these were called Multifract HS-40, HS-42, and HS-45, the numbers indicating the number of vials in the headspace carousel. A later version was called HS-100, and this was mounted on a Sigma 2000 gas chromatograph allowing automated analysis of up to 100 specimens. The AutoSystem XL GC is the most recent development

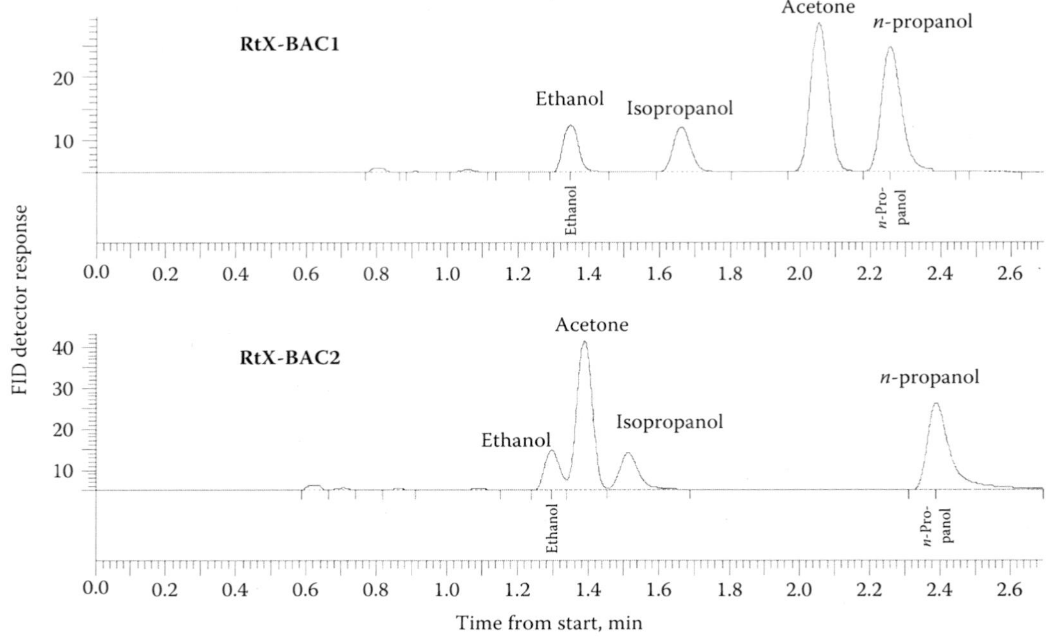

Figure 2.6 Gas chromatographic traces obtained from analysis of an aqueous mixture containing several volatile substances often encountered in forensic blood samples. The analysis was done by headspace gas chromatography using commercially available capillary columns (Restek Corporation) RtX-BAC1 and RtX-BAC2.

and works together with TurboMatrix 110 headspace sampler. This arrangement permits overnight batch analysis of up to 110 specimens in a single run.

Packed, wide-bore, and capillary columns are feasible together with headspace gas chromatography, and for high resolution work, such as when complex mixtures are being analyzed, capillary columns are essential.[93,94] Traditional packed columns made of glass or stainless steel are, however, more robust and are still widely used in some laboratories for routine blood-alcohol analysis. Figure 2.6 gives examples of gas chromatographic traces obtained by headspace analysis of an aqueous mixture of volatiles commonly encountered in forensic toxicology. The column was made of capillary glass designed and marketed especially for blood-alcohol analysis (RtX-BAC1 and RtX-BAC2) and purchased from Restek Corporation (Bellefonte, PA, U.S.A.). The ethanol response is well resolved from potential interfering substances in an isothermal run lasting for 2 min. Table 2.5 gives retention times relative to *n*-propanol as internal standard for a wider range of low molecular volatiles under normal HS-GC conditions.

Sampling and analysis of the vapor in equilibrium with the blood specimen has the advantage that non-volatile constituents of the biological matrix (fats, proteins, etc.) do not clog the syringe or the column packing material. Sensitivity of the assay can be enhanced and matrix effects eliminated in another way, namely, by saturating the blood samples and aqueous ethanol standards with an inorganic salt such as NaCl, K_2CO_3, or Na_2SO_4, e.g., 0.5 mL blood + 1 g salt.[86,89] A salting-out technique is useful when trace concentrations of volatiles are analyzed in blood such as endogenous alcohols or alcoholic beverage congeners.[95–98]

More recently, the headspace vapor in equilibrium with blood or other body fluid can also be removed and transferred to a GC instrument with a solid phase micro-extraction probe.[99–100] The needle of the probe contains a porous polymer material and this is inserted into the headspace vapor to attain equilibrium with any volatiles in the flask. The probe is then withdrawn from the vial and introduced into the heated injection port of the gas chromatograph. This sampling technique

Table 2.5 Relative Retention Times (min) for Analysis of Volatile Compounds by Headspace Gas Chromatography Using Two Different Stationary Phases

Substance	Stationary Phase RtX-BAC1	Stationary Phase RtX-BAC2
Acetaldehyde	0.53	0.34
Acetone	0.91	0.57
Acetonitrile	0.91	0.81
1-Butanol	2.05	2.20
2-Butanone	1.37	1.04
t-Butanol	0.88	0.69
Ethanol	0.59	0.53
Isopropanol	0.74	0.62
Methanol	0.47	0.40

Note: Times are relative to *n*-propanol as internal standard and all determinations were made with a Perkin-Elmer AutoSystem XL gas chromatograph in isothermal mode and a TurboMatrix 110 headspace sampler.

is well suited for analysis of a wide range of volatile agents such as aliphatic and aromatic hydrocarbons as well as many water-soluble alcohols and ketones.[99]

Another modification of the standard headspace analysis involved inclusion of a cryofocusing step prior to GC analysis with a liquid nitrogen freeze trap.[101] This serves to concentrate the specimen prior to chromatographic analysis of volatiles with a capillary or wide-bore column. This was the approach used to measure trace amounts of endogenous volatile alcohols in blood samples or to establish the congener profile in blood after ingesting different kinds of alcoholic beverages.[97,98,101]

Gas chromatographic methods of analysis have the unique advantage that they combine a qualitative screening analysis of the components of a mixture based on their relative retention time after injection to appearance of the peak with a simultaneous quantitative analysis by measuring the detector response as reflected in the height or area of the resulting peak response.[102] Several comprehensive reviews have dealt with forensic applications of gas chromatography including applications for blood-alcohol analysis.[14,17,86,103,104] One of these reviews looked at more general applications of headspace analysis when applied to biological specimens for determination of organic volatile substances, including alcohols.[104]

In forensic work, it is advisable to use two different column packing materials, thus furnishing different retention times for ethanol and other volatiles that might be encountered in forensic blood samples. This becomes important whenever blood or tissue samples are putrefied and therefore might contain interfering substances with the same retention times as ethanol on a single stationary phase. The risk of obtaining coincident retention times on two or more stationary phases is reduced considerably. Otherwise, two different methodologies such as GC and chemical oxidation or GC and enzymatic oxidation could be used to analyze duplicate aliquots from the same blood sample.[105] HS-GC with two different detectors (flame ionization and electron capture) has been used to screen biological fluids for a large number of volatiles.[106] Such a dual-detector system was recommended for use in clinical toxicology to aid in the diagnosis of acute poisoning when a host of unknown substances might be responsible for the patient's condition.[107]

2.3.4 Other Methods

A multitude of other analytical methods exist for blood-alcohol analysis, but none of these can match HS-GC, which is considered the gold standard in forensic and clinical toxicology laboratories. Instead of a flame ionization detector, electrochemical sensing[108] or a metal oxide semiconductor device[109] has been applied to headspace analysis of blood samples. However, lack of a chromatographic separation step meant that neither of these procedures was sufficiently selective when interfering substances might be present. Some authorities recommend making a rapid screening of biological samples by ADH methods to eliminate those that do not contain any alcohol.[66] All positives are later run by the usual headspace analysis and gas chromatography. This same approach proved suitable for measuring the strength of alcoholic beverages and gave good results compared with a standard gas chromatographic method.[110]

Several novel methods for alcohol analysis make use of biosensors prepared from immobilized enzymes or bioelectrodes, and these have found applications in clinical chemistry laboratories.[111–115] The end point of the enzymatic reactions can be monitored either by amperometric, colorimetric, or spectrophotometric methods.[113,114] The enzyme alcohol oxidase has attracted attention for analysis of alcohol in body fluids and gives reasonably good semiquantitative results.[115–118] These systems are similar in principle to measuring blood glucose with a glucose oxidase electrode and open the possibility for self-testing applications such as the glucose dipstick technology.

Fourier transform infrared spectrometry (FTIR) was recently applied to the determination of alcohol in beer,[119] and when a purge-and-trap capillary GC separation stage was included, FTIR could also be adopted to measure a wide range of low-molecular-weight volatiles including ethanol.[120] A method based on proton nuclear magnetic resonance spectroscopy proved suitable for use in pharmacokinetic studies to analyze ethanol, acetone, and isopropanol in plasma samples.[121,122]

In clinical and emergency medicine, depression of freezing point (osmometry) has a long history as a screening test for certain pathological conditions.[123,124] Diabetes mellitus and uremia are often associated with abnormally high concentrations of plasma-glucose and plasma-urea, respectively. These common conditions cause discrepancies between the osmolality expected from the inorganic ions Na^+ and K^+ and the values measured by depression of the freezing point. Dedicated equipment for osmometry is available at most hospital laboratories and only about 0.2 mL of plasma is needed to measure freezing-point depression. Moreover, the method is nondestructive, which means that the same specimen of plasma can be used later for making a confirmatory toxicological analysis if necessary.[124]

The osmolal gap is defined as the difference between the measured serum osmolality determined by freezing point depression and the calculated serum molarity, from known osmotically active substances in the serum specimen (sodium, potassium urea, glucose, and ethanol).[40] Indeed, in emergency medicine, ethanol is the commonest cause of finding a high serum osmolality.[124–126] Ethanol carries an appreciable osmotic effect because of its low molecular weight (46.05), high solubility in water, and the fact that large quantities are ingested to produce gross intoxication.[126] Finding a normal osmolal gap speaks against the presence of a high concentration of ethanol but a high osmolal gap does not necessarily rule out that this was caused by ethanol. Other toxic solvents in serum such as acetone, methanol, isopropanol, and ethylene glycol will increase the serum or plasma osmolality as revealed by freezing point depression.[126] The principal limitation of using osmolal gap as a rapid test for high serum ethanol is the lack of selectivity because other toxic alcohols and non-electrolytes if present will be falsely reported as ethanol. Nevertheless, papers continue to be published dealing with the principles and practice of freezing point osmometry in emergency toxicology.[127]

Considerable interest has developed in point-of-care or near patient testing and in this connection non-invasive methods are preferred. Near-infrared spectrometry is a technique with huge potential for non-invasive analysis of various substances (e.g., tissue glucose) and more recently also tissue ethanol.[128,129] A light with fixed wavelength is beamed through a subject's fingertip or arm and after processing the absorption bands of the emitted light into their specific wavelengths various constituents in the tissue water can be identified and for some substances a quantitative analysis is possible. However, disentangling the signals of interest from the background noise generated by other biological molecules has proved a challenging problem. Progress is rapidly being made with the aid of sophisticated computer-aided pattern recognition techniques. Near-infrared spectroscopy has already been successfully applied to the analysis of glucose,[130] and a recent publication described the application of a similar technique for analysis of alcohol in tissue in a completely non-invasive way.[129] The future of such technology in clinical and forensic work remains to be seen.

Table 2.6 Classification of Currently Available Instruments for Breath-Alcohol Analysis According to the Main Area of Application and the Analytical Principle Used

Instrument	Main Area of Application	Analytical Principle
Alcolmeter	Roadside screening of motorist, also in the workplace and at hospital casualty departments	Electrochemical oxidation (fuel cell)
Alco-Sensor*	Roadside screening of motorist, also in the workplace and at hospital casualty departments	Electrochemical oxidation (fuel cell)
Alcotest*	Roadside screening of motorist, also in the workplace and at hospital casualty departments	Electrochemical oxidation (fuel cell)
Lifeloc	Roadside screening of motorist, also in the workplace and at hospital casualty departments	Electrochemical oxidation (fuel cell)
Alcotest 7110	Evidence for prosecution of drinking drivers	Electrochemical oxidation and infrared (9.5 μm)
Intoxilyzer 8000	Evidence for prosecution of drinking drivers	Infrared analysis at 3.4 μm and 9.5 μm
BAC Datamaster	Evidence for prosecution of drinking drivers	Infrared analysis at three wavelengths close to 3.4 μm
Intoximeter EC/IR	Evidence for prosecution of drinking drivers	Electrochemical oxidation

* Also being used for roadside evidential testing in some U.S. states.

The feasibility of combining gas chromatography (GC) to separate the volatile components in a mixture and mass spectrometry (MS) as the detector was demonstrated many years ago.[131] GC-MS provides an unequivocal qualitative analysis of ethanol from its three major mass fragments: *m/z* 31 (base peak), *m/z* 45, and *m/z* 46 (molecular ion).[132] Isotope-dilution GC-MS is of course feasible if instead of *n*-propanol as internal standard d_5-ethnaol is used ($C^2H_3{}^2H_2OH$).[133] Selected ion monitoring and deuterium-labeled ethanol were also used to distinguish between ethanol formed post-mortem by the action of bacteria on blood glucose using an animal model.[134,135] In a clinical pharmacokinetic study, unlabeled ethanol was administered mixed with its deuterium-labeled analogue to investigate the bioavailability of ethanol and the role of first-pass metabolism in the gut.[136]

2.4 BREATH-ALCOHOL ANALYSIS

The smell of alcohol on the breath of a drinker has always been recognized as a sign of excess alcohol consumption. Quantitative studies demonstrate that only about 1 to 2% of the alcohol ingested is expelled unchanged in the breath. Breath-alcohol instruments were developed to provide a fast and non-invasive way of monitoring alcohol concentration in the blood. A large body of literature has dealt with the principles and practice of breath-alcohol analysis and the associated technology for applications in research, clinical practice, and law enforcement.[137–142] Analysis of a person's expired air furnishes an indirect way of monitoring volatile endogenous substances in the pulmonary blood and this approach has many interesting applications in clinical and diagnostic medicine.[143,144] However, the main application of breath-alcohol instruments is in the field of traffic-law enforcement for testing drunk drivers and more recently also for workplace alcohol testing.[137,141,142] Two categories of instrument for breath-alcohol analysis can be distinguished depending on whether the results are intended as a qualitative screening test for alcohol or as binding evidence for prosecution of drunk drivers (Table 2.6).

2.4.1 Handheld Screening Instruments

Various handheld devices are available for roadside pre-arrest screening of drinking drivers to indicate whether a certain threshold concentration of alcohol has been surpassed.[141,142] Such screening tests are usually conducted at the roadside, and for evidential purposes, a more controlled breath-alcohol analysis is usually mandatory. The instruments for evidential purposes are larger, more sophisticated, and include ways to check accurate calibration, analyze alcohol-free room air, and produce a printed record of the results. In short, they furnish a quantitative analysis of BrAC and serve as binding evidence for prosecuting drunk drivers.[145,146] Breath-alcohol instruments have also found applications in clinical pharmacokinetic studies of ethanol and drug-alcohol interactions.[147,148] Handheld breath-alcohol analyzers are also very practical for use in emergency medicine as a quick and easy way to monitor whether a patient's behavior and signs and symptoms of impairment can be attributed to alcohol influence.[149–151]

The analytical principles for measuring ethanol in the exhaled air depend in part on the area of application, that is, whether results are intended for qualitative screening or evidential purposes.[148] Most handheld screening devices incorporate electrochemical "fuel-cell" sensors that oxidize ethanol to acetaldehyde and in the process produce free electrons. The electric current generated is directly proportional to the amount of ethanol consumed by the cell (Table 2.6). Acetone, which is the most abundant endogenous volatile exhaled in breath, is not oxidized at the electrode surface so elevated concentrations of this ketone (e.g., in untreated diabetes) will not cause a false-positive response.[152] However, if high concentrations of methanol or isopropanol are present in exhaled breath, these also undergo electrochemical oxidation although at different rates compared with ethanol.[151] Care is needed when test results with fuel-cell instruments are interpreted because isopropanol, under some circumstances, can be formed in the body by reduction of endogenous acetone.[153] The concentration of acetone in blood reaches abnormally high levels during food deprivation, prolonged fasting (dieting), or during diabetic ketoacidosis.[152]

2.4.2 Evidential Breath-Testing Instruments

Most of the evidential breath-testing instruments used today identify and measure the concentration of alcohol by its absorption of infrared energy at wavelengths of 3.4 or 9.5 µm, which corresponds to the C–H and C–O vibration stretching in the ethanol molecules, respectively (Table 2.6).[17,154–156] Selectivity for identifying ethanol is enhanced by combining infrared absorption at 9.5 µm and electrochemical oxidation within the same unit and the Alcotest 7110 features this dual-sensor technique.[141,142] Another example from the latest generation of breath-test instruments is the Intoxilyzer 8000, which makes use of infrared wavelengths at 3.4 and 9.5 µm for identification and analysis of ethanol. This reduces considerably the risk of other breath volatiles, if any exist, being reported as ethanol.

Modern evidential breath-alcohol instruments are equipped with microprocessors that control the entire breath-test sequence, including the exhaled volume and alcohol concentration, and the exact shape of the BrAC–time profile is monitored and stored in the computer. Examples of exhalation profiles for two individuals tested with a state-of-the-art evidential breath-alcohol analyzer are given in Figure 2.7. One notices a rapid rise in the concentration of exhaled ethanol, and after just a few seconds of starting the exhalation, 70% of the final value is reached. Thereafter, the BrAC continues to increase although this occurs much more gradually until the person reaches the end of the exhalation after 9 to 10 s. Note that a BrAC plateau is never reached.

The rules and regulations governing evidential breath-alcohol tests stipulate the need for an observation and deprivation period of at least 15 min before starting the test.[157–159] Immediately after finishing a drink and for some time afterwards, the concentration of alcohol in the mucous surfaces of the mouth will be higher than that expected from the coexisting BAC. Time is needed for the alcohol to dissipate, and many studies have shown that this takes 15 to 20 min even after gargling with whisky or mouthwash.[160,161] The latest generation of breath-alcohol analyzers is equipped with algorithms that monitor the shape of the BrAC exhalation profile to help disclose abnormalities that might be caused by mouth alcohol. Besides the problem with alcohol in the mouth from a recent drink, the operator of the breath instrument should also ensure the person

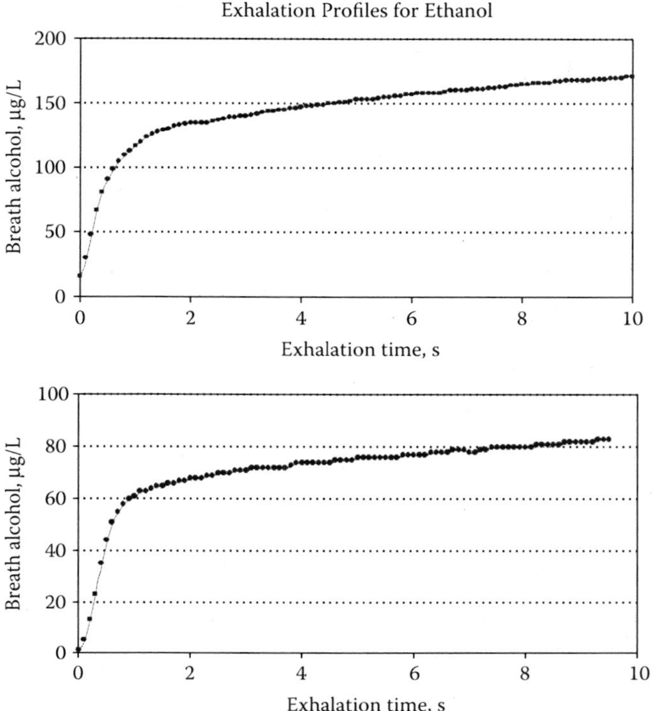

Figure 2.7 Breath-alcohol profiles during a prolonged exhalation by two test subjects.

does not hiccup, burp, belch, or regurgitate stomach contents just before testing. This might result in a more dangerous form of mouth alcohol that is not so easily detected by monitoring the slope of the exhalation profile.[162]

Reporting results of breath-alcohol analysis can be a bit confusing and depends on whether the testing was done for clinical or traffic law enforcement purposes. In hospitals, it is standard practice to translate the measured BrAC into the presumed concentration in venous blood, which requires use of a calibration factor. This is referred to as the blood/breath ratio and the instrument gives results directly in terms of BAC derived as [BrAC × ratio = BAC]. The value of the blood/breath ratio is generally taken to be 2100:1 (U.S. and Canada), although values of 2300:1 (U.K., Ireland, and Holland) and 2000:1 (Germany, France, Spain) are accepted. For legal purposes the BrAC is not converted to BAC but instead as g/210 L breath in U.S., so the 2100:1 blood/breath ratio is affirmed by statute.[159,160] However, many blood-alcohol and breath-alcohol comparisons show that a factor of 2100:1 gives a generous margin of safety to the individual.

Direct comparisons have shown that the concentration of alcohol in venous blood is about 10 to 15% higher compared with BrAC (× 2100) when samples are taken in the post-absorptive phase.[153] A closer agreement between the two methods of measurement is obtained using a 2300:1 factor for calibration purposes.[154] This is illustrated in Figure 2.8, which compares mean concentration-time profiles in venous blood and end-exhaled breath in ten subjects after they drank 0.4 g ethanol per kg body weight at about 2 h after their last meal.[156] The breath-alcohol instrument in this study was an Intoxilyzer 5000, which has been widely used for legal purposes in the U.S. and elsewhere.[153,155,163]

2.4.3 Blood/Breath Ratio of Alcohol

Studies have shown that the BAC/BrAC ratio of alcohol changes as a function of the time after drinking depending on whether tests were made on the absorption or the post-absorptive part of

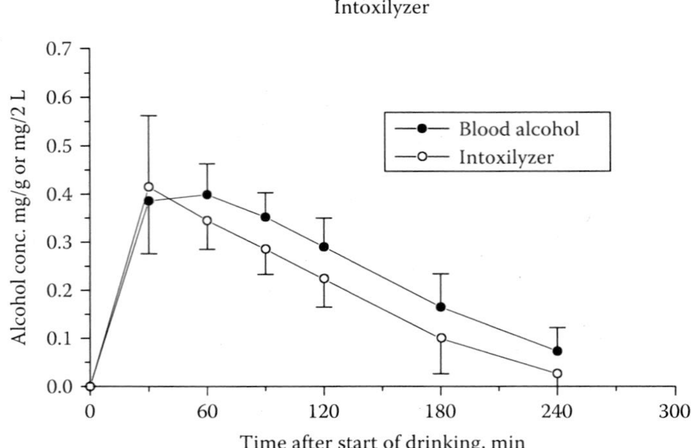

Figure 2.8 Relationship between the concentrations of ethanol in venous blood and end-expired breath determined with Intoxilyzer 5000, a quantitative infrared analyzer.

the alcohol curve.[164–166] The blood-to-breath alcohol ratio tends to be less than 2100:1 (1800 to 2000) during the absorption phase and greater than 2100:1 (2200 to 2400) in the post-absorptive phase. One reason for this temporal variation stems from the fact that venous instead of arterial blood was used for determination of alcohol.[167,168] The alcohol concentration in pulmonary blood reaching the air sacs (alveoli) of the lungs runs closer to the concentration in arterial blood transporting alcohol to tissue water compared with venous blood returning blood to the heart.

Figure 2.9 shows mean concentration–time profiles of alcohol derived from blood samples drawn from indwelling catheters in a radial artery at the wrist and a cubital vein at the elbow on the same arm.[167] The volunteers were healthy men who drank a moderate bolus dose of ethanol (0.6 g/kg body weight) 5 to 15 min before providing specimens of venous and arterial blood nearly simultaneously at various times after drinking ended. The concentration of alcohol in arterial blood samples exceeded venous blood samples during the first 90 min post-dosing, whereas at all later

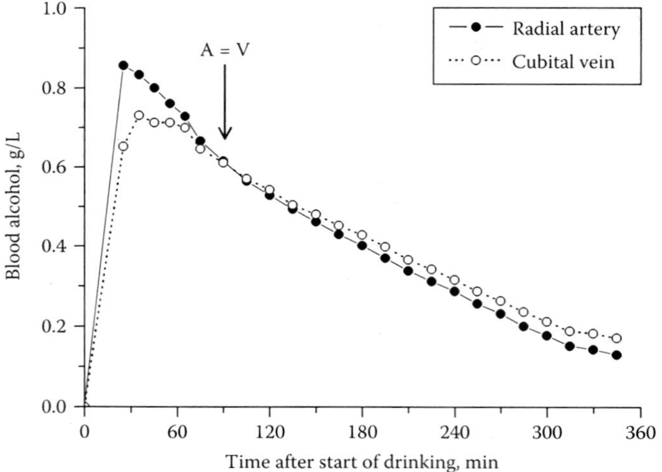

Figure 2.9 Mean blood-alcohol curves comparing the concentrations of ethanol in blood from a radial artery and a cubital vein on the same arm. The curves depict nine healthy men who drank 0.6 g ethanol per kg body weight in 2 to 15 min. Standard error bars are omitted for clarity.

times the venous blood was slightly higher than arterial blood. The arterial and venous concentrations were the same at only one time point at about 90 min post-drinking.

Note that for clinical applications, the breath-alcohol instruments are calibrated to estimate the alcohol concentration in whole blood and not the concentration in plasma or serum. This is often overlooked by clinicians who fail to appreciate the difference between whole blood and plasma concentrations of alcohol. To derive the concentration of alcohol in plasma or serum indirectly by analysis of breath, the instrument would need to be calibrated with a plasma/breath factor of about 2600:1 because whole blood contains about 15% less alcohol than the same volume of plasma or serum (Table 2.3).

When breath-alcohol testing is used for traffic law enforcement the results are almost always reported as the concentration of alcohol in the breath analyzed without considering the person's BAC. This avoids making any assumptions about the blood/breath ratio and its variability between and within individuals. Statutory limits for driving in many countries are therefore written in terms of threshold BAC and BrAC depending on the specimen analyzed.[166] Evidential breath-alcohol tests should be done in duplicate and the protocol should include analysis of room air blank and a known strength air–alcohol mixture as a control standard.[169,170] Results from all breath and control tests along with the date and time of testing as well as positive identification of the suspect should be available as a print-out and also stored online for later downloading to a central computer network.[169,170]

2.5 QUALITY ASSURANCE ASPECTS OF ALCOHOL ANALYSIS

Much has been written about quality assurance of clinical laboratory analysis including concepts such as precision, accuracy, linearity, recovery, sensitivity, and limits of detection and quantitation of the method.[171] In addition, when results are used as evidence in criminal and civil litigation, the chain-of-custody record of the specimens is extremely important to document. This chain must be maintained intact from the moment of sampling to the moment results are reported, and each person involved in handling, transport, analysis, and storage of the specimen must be traceable from the written records. The entire analytical procedure including the actual chromatographic traces as well as proof that the instrument was properly calibrated on the day the blood specimens were analyzed might need to be verified at a later date. Participation in nationally or internationally recognized external proficiency tests is another essential element of quality assurance of laboratory results.[172]

The important concepts in development and validation of analytical methods include accuracy, precision, linearity, range, specificity/selectivity, limit of detection, and quantitation, and these terms are explained in brief below.[170,171]

Accuracy is defined as the closeness of agreement of the analytical result with a known reference or assigned value. Accuracy is therefore a measure of the exactness of the method. The difference if any between the true value and the value found is called analytical bias.

Precision describes the magnitude of random error as reflected in the closeness of agreement of multiple determinations on the same specimen. If the replicate measurements are made within the same analytical run, the calculated precision estimated is referred to as repeatability of the method. When replicates are made in different runs or at different laboratories by different people, then the analytical precision is referred to as reproducibility of the method. In mathematical terms, precision is reflected in the standard deviation (SD) of at least ten replicate measurements at a certain concentration of the target analyte. For obvious reasons, reproducibility can be expected to be poorer than repeatability. For blood-ethanol assay, the SD tends to increase as concentration in the specimen increases, and therefore precision is often expressed as a coefficient of variation (SD/mean) $\times$ 100, which tends to decrease slightly as the concentration of ethanol increases.

The terms **sensitivity** and **specificity** are often used interchangeably and they generally refer to a method's ability to distinguish between different analytes in a mixture and allow a response for just the substance of interest, that is, without interference from potential endogenous or exogenous compounds.

Linearity is the ability of a method to give results that are directly proportional to concentration of the target analyte or can be expressed by a well-defined mathematical transformation and shown to be a function of concentration over a certain range.

The **range** is the interval between the upper and lower concentrations of analyte that can be determined with sufficient accuracy, precision, and linearity.

The **limit of detection** (LOD) is the lowest concentration of analyte in a specimen that can be distinguished from background signals or instrument noise. Mathematically LOD is defined as 3 × the SD obtained from analysis of a blank specimen.

The **limit of quantitation** (LOQ) is the concentration of analyte in a sample that can be determined with certainty. Mathematically LOD is defined as 10 × the SD obtained from analysis of a blank specimen.

The most important features of pre-analytical, analytical, and post-analytical aspects of blood-alcohol analysis are presented below.

2.5.1 Pre-Analytical Factors

The subject or patient should be informed of the reason for taking a blood sample and, when necessary, informed written consent obtained. The equipment used for drawing blood is normally an evacuated tube (5- or 10-mL Vacutainer tubes) with sterile needle attachment. The blood is taken from an antecubital vein, and if necessary a tourniquet is applied to make it easier to visualize a suitable vein. The volume of blood drawn depends on the requirements and number of substances to be analyzed, whether only ethanol or ethanol and other drugs of abuse. Even though in the case of ethanol only a few hundred microliters of blood are needed for each assay, the Vacutainer tubes should be filled with as much blood as possible. Sufficient blood should be available to allow making several determinations of the ethanol concentration and any retesting that might be necessary as well as common drugs of abuse. The specimen tubes should be gently inverted a few times immediately after collection to facilitate mixing and dissolution of the chemical preservatives; sodium fluoride (10 mg/mL) to inhibit the activity of various enzymes, microorganisms, and yeasts; and potassium oxalate (5 mg/mL) as an anticoagulant. The tubes of blood should be labeled with the person's name and the date and time of sampling, and the name of the person who took the sample should also be recorded. The Vacutainer tubes containing blood should be sealed in such a way as to prevent unauthorized handling or tampering; special adhesive paper strips are available for this purpose. The blood samples and other relevant paperwork should then be secured with tape so that any deliberate manipulating or adulteration is easily detected by laboratory personnel after shipment. After taking the samples, the tubes of blood should be stored in a cold room before being sent to the laboratory by express courier mail service.

The question of a deficient blood volume in the Vacutainer tubes sent for analysis of alcohol and the influence this might have on accuracy of HS-GC analysis was recently investigated in two separate publications.[173,174] It was alleged that a very small volume of blood meant an excess amount of sodium fluoride and that this increased the ethanol concentration in the headspace flask by a salting-out effect. This led to an acquittal in a drunken driving case trial in the court of appeals in the U.K., and the prosecution failed to rebut the argument with expert testimony. The new studies showed that an abnormally high concentration of NaF increased not only the concentration of ethanol in the headspace, but also the concentration of the *n*-propanol internal standard. Because the ratio of responses EtOH/PrOH is used for quantitative analysis, the resulting ethanol concentration should be unchanged. In reality, it was shown that *n*-propanol was salted-out slightly more effectively than ethanol with the net result that the blood-ethanol concentration reported was lower than expected and was thus to the suspect's benefit.

Although pre-analytical factors are probably more important to consider when endogenous substances are analyzed such as in clinical chemistry laboratories, a standardized sampling protocol is also important for forensic blood-alcohol analysis.[175,176] Two tubes of blood should be drawn in rapid succession and the skin cleaned with soap and water and not with an organic solvent such as ether, isopropanol, or ethanol. Obviously, the blood samples should not be taken from veins into which intravenous fluids are being administered.[177] This kind of emergency treatment is often given as a first-aid to counteract shock or trauma as a result of involvement in traffic accidents. The blood samples should be taken only by trained personnel such as a phlebotomist, registered nurse, or physician.

2.5.2 Analytical Factors

The blood specimens must be carefully inspected when they arrive at the laboratory, being noted whether the seals on the package as well as the individual tubes of blood are intact. If the specimen seems unusually dilute or if there are blood clots, then this should be noted. There are two ways to deal with clotted samples: (1) either centrifuge the specimen and use an aliquot of the supernatant for analysis and then report the results as a serum concentration or (2) homogenize the clot and analyze an aliquot of the hemolyzed blood specimen. Details of any mishaps occurring during transport of specimens such as breakage of the packaging, leakage of blood, etc., as well as the date and time of arrival, should be recorded. Information written on the Vacutainer tubes should be compared with other documentation to ensure the suspect's name and the date and time of sampling are correct. The same unique identification number or barcode should be added to all paperwork and biological specimens received and this number used to monitor passage of the specimens through the laboratory. Ensure that the erythrocytes and plasma fractions are adequately mixed before removing aliquots of whole blood for analysis of ethanol. Replicate determinations can be made with different chromatographic systems and preferably by different technicians working independently. Any unidentified peaks on the gas chromatograms should be noted because these might indicate the presence of other volatiles in the blood sample.

2.5.3 Post-Analytical Factors

Quality assurance of individual results can be controlled by looking at critical differences (range) between replicate determinations.[170] The size of the difference will be larger, the higher the concentration of ethanol in the blood specimen because precision tends to decrease with an increase in the concentration of ethanol. Control charts offer a useful way to monitor day-to-day performance in the laboratory; one chart is used to depict random errors or precision and another chart to show systematic errors or bias derived by analysis of known strength standards together with unknowns.[178,179] These charts make it easy to detect sudden deterioration in analytical performance as shown by the scatter of individual values and the number of outliers.[178]

The rate of loss of alcohol during storage needs to be established under refrigerated conditions (+4°C) and also when specimens are kept deep frozen.[180–182] If necessary, corrections can be applied to blood specimen reanalyzed after prolonged periods of storage. In our experience, the blood-ethanol concentration decreases only slightly during storage at 4°C at a rate of 1 to 2 mg/dL per month.[148] Chromatographic traces and other evidence corroborating the analytical results such as calibration plots or response factors all need to be carefully labeled and stored in fireproof cabinets. Today it is virtually mandatory that a forensic science or toxicology laboratory be accredited for the tasks they perform.[183,184] Participation in external proficiency trials of analytical performance is also a mandatory requirement to build confidence in the analytical reports.[172,185] When analytical results are used for criminal prosecution, information of the kind discussed above must be open to discovery and needs to be available for scrutiny.

Table 2.7 Results of a Declared Interlaboratory Proficiency Test of Blood-Alcohol Analysis at Specialist Forensic Toxicology Laboratories in the Nordic Countries

Laboratory	Blood-1	Blood-2	Blood-3	Blood-4	Blood-5	Blood-6
1	0.46	1.01	2.15	1.62	0.74	1.75
2	0.47	1.01	2.27	1.70	0.78	1.83
3	0.46	1.01	2.26	1.67	0.77	1.81
4	0.47	1.00	2.17	1.66	0.78	1.81
5	0.48	1.01	2.15	1.66	0.78	1.79
Mean	0.47	1.01	2.20	1.66	0.77	1.80
SD	0.008	0.005	0.060	0.029	0.017	0.030
CV%	1.7%	0.49%	2.7%	1.7%	2.2%	1.7%

Notes: Venous blood samples were taken from apprehended drinking drivers in sterile tubes containing sodium fluoride and potassium oxalate as preservatives. The blood was portioned out into other tubes and sent to participating laboratories as a declared proficiency trial. As seen, the CVs between laboratories were always less than 3% and the corresponding within-laboratory CVs were mostly less than 1% (data not shown).

2.5.4 Interlaboratory Proficiency Tests

Two papers looked at the results from interlaboratory proficiency tests of blood-alcohol analysis at clinical chemistry laboratories.[172,185] In one study originating from Sweden, all participants were clinical chemistry laboratories and all used gas chromatographic analysis of plasma-ethanol. The coefficients of variation (CVs) between laboratories were within the range 10 to 17%.[172] In a similar study among U.K. laboratories, the corresponding CVs depended in part on the kind of methodology used for determination of alcohol and immunoassays generally performed worse than gas chromatographic methods (liquid injection and headspace technique) and the CVs ranged from 8 to 20%.[185] It should be noted that determination of toxicological substances such as ethanol in blood is not the primary concern of clinical chemistry laboratories.

Table 2.7 presents results from an interlaboratory comparison of blood-alcohol analysis at specialist forensic toxicology laboratories in the Nordic countries (Denmark, Finland, Iceland, Norway, and Sweden). All participants used headspace gas chromatography for the determinations and the blood samples were obtained from apprehended drinking drivers. The CV between laboratories was always less than 3% regardless of the concentration of alcohol present, which testifies to highly reproducible analytical work. The corresponding CVs within laboratories were mostly 1% or less based on three to six determinations per sample. If the overall mean BAC in each sample is taken as the target value, then all laboratories showed accuracy to within ±5% of the attributed concentration.

2.6 FATE OF ALCOHOL IN THE BODY

Ethanol is a small polar molecule with a low molecular weight (46.07) and carries weak charge (see Table 2.2), which facilitates easy passage through biological membranes.[186,187] After ingestion, absorption of ethanol starts already in the stomach, but this process occurs much faster from the upper part of the small intestine where the available surface area is much larger owing to the presence of microscopic villi covering the mucosal cells. Both the rate and extent of absorption are delayed if there is food in the stomach before drinking.[188,189] The blood that drains the gastrointestinal tract leads to the portal vein where any alcohol present must pass through the liver, and then via the hepatic vein on to the heart and the systemic circulation. After passage through the liver, ethanol distributes uniformly throughout all body fluids and tissue without binding to plasma proteins. Indeed, it is possible to determine total body water by the ethanol dilution method.[187]

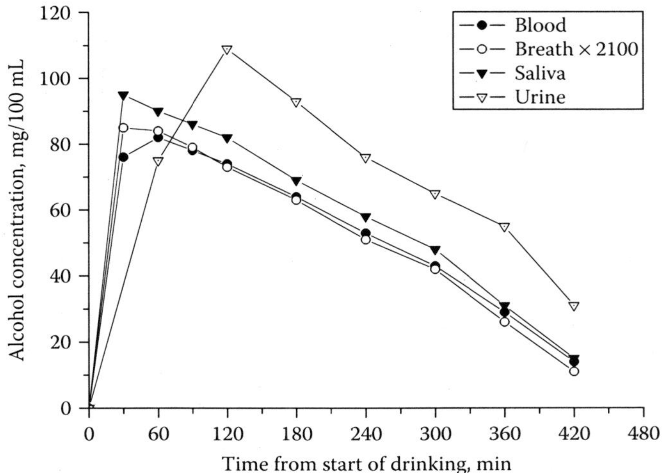

Figure 2.10 Mean concentration–time profiles of ethanol in venous blood, end-expired breath (× 2100), saliva, and urine from experiments with healthy men who ingested 0.68 g ethanol per kg body weight as neat whisky in the morning after an overnight fast. Standard error bars are omitted for clarity.

The peak BAC reached after drinking (C_{max}) and the time required to reach the peak (t_{max}) vary widely from person to person and depend on many factors.[187] After 48 healthy male volunteers drank 0.68 g ethanol/kg body weight as neat whisky on an empty stomach, the peak concentration in capillary (fingertip) blood was reached at exactly 10, 40, 70, and 100 min after the end of drinking for 23, 14, 8, and 3 subjects, respectively.[190] The amount of alcohol consumed, the rate of drinking, the dosage form (beer, wine, spirits, cocktails), and most importantly the rate of gastric emptying will influence the speed of ethanol absorption.[187] The concentrations of ethanol in body fluids and tissues after reaching equilibration depend primarily on the water contents of these fluids and tissues and the ratio of blood flow to tissue mass.[187] Figure 2.10 shows the mean concentration–time profiles of ethanol in blood, breath, urine, and saliva obtained in experiments with healthy male volunteers who drank 0.68 g/kg as neat whisky in 20 min after an overnight fast.[187,191] Note that the concentrations in breath have been multiplied by an assumed average blood/breath ratio of 2100:1.

The bulk of the dose of alcohol (95 to 98%) is eliminated by oxidative metabolism, which occurs via enzymatic reactions in the liver catalyzed by class I enzymes of alcohol dehydrogenase (ADH).[186,187] Between 2 and 5% of the dose is excreted unchanged in breath, urine, and sweat and a very small fraction (~0.1%) is conjugated with glucuronic acid and removed via the kidney.[192] Small amounts of alcohol are thought to undergo pre-systemic oxidation in the gastric mucosa or the liver or both organs but the practical significance of first-pass metabolism (FPM) is dose dependent and is not easy to quantify.[193]

At moderate BAC (>60 mg/dL), the microsomal enzymes (P4502E1), which have a higher k_m for oxidation of ethanol (60 to 80 mg/dL) compared with ADH (k_m = 2 to 5 mg/dL), become engaged in the metabolism of ethanol.[194–196] The P450 enzymes are also involved in the metabolism of many drugs and environmental chemicals, which raises the potential for drug–alcohol interactions, which might explain the toxicity of ethanol in heavy drinkers and alcoholics.[197–202] Moreover, the activity of P4502E1 enzymes increases after a period of continuous heavy drinking owing to a faster *de novo* synthesis of the enzyme and metabolic tolerance develops as reflected in twofold to threefold faster rates of elimination of alcohol from the bloodstream in alcoholics undergoing detoxification.[200–204]

The detrimental effects of ethanol on performance and behavior are complex and involve interaction with the membrane receptors in the brain associated with the inhibitory neurotransmitters glutamate and gamma aminobutyric acid (GABA).[205–208] The behavioral effects of ethanol are dose-dependent and after drinking small amounts the individual relaxes, experiences mild euphoria, and becomes more talkative. As drinking continues and the blood-ethanol concentration increases toward 150 to 200 mg/dL, impairment of body functioning becomes pronounced. Many of the pharmacological effects of ethanol can be explained by an altered flux of ions through the chloride channel activated by the neurotransmitter GABA.[206] The link between ethanol impairment and neurotransmission at the $GABA_A$ receptor also helps to explain observations about cross-tolerance with other classes of depressant drugs like benzodiazepines and barbiturates, which also bind to the $GABA_A$ receptor complex to open a chloride ion-channel to alter brain functioning.[200]

Although there is a reasonably good correlation between degree of ethanol-induced impairment and the person's BAC, there are large variations in response at the same BAC in different individuals who drink the same amount of alcohol within the same time frame. The reasons for this are twofold; first, larger people tend to have more body water so the same dose of alcohol enters a larger volume resulting in lower BAC compared with lighter people with less body water. This phenomenon is known as consumption tolerance and stems from variations in body weight and the relative amount of adipose tissue, which is influenced by age, gender, and ethnicity.[209,210] The second reason for interindividual differences in ethanol-induced performance decrement is called concentration tolerance, which is linked to a gradual habituation of brain cells to the presence of alcohol during repeated exposure to the drug.[210,211] Besides the development of acute tolerance (Mellanby effect), which appears during a single exposure (see Chapter 1), a chronic tolerance develops after a period of continuous heavy drinking. Among the mechanisms accounting for chronic tolerance are long-term changes in the composition of cell membranes particularly, the cholesterol content, the structure of the fatty acids, and also the arrangement of proteins and phospholipids making up the lipid bilayer.[210,211]

In occasional drinkers, the impairment effects of ethanol appear gradually, becoming more exaggerated as BAC increases. The various clinical signs and symptoms of intoxication are usually classified as a function of BAC from sober to dead drunk as was first proposed by Bogen.[212] This scheme has subsequently been developed further and improved upon by others. For example, at a BAC of 10 to 30 mg/dL alterations in a person's performance and behavior are insignificant and can only be discerned using highly specialized tests such as divided attention tasks. Between 30 and 60 mg/dL, most people experience euphoria, becoming more talkative and sociable owing to disinhibition. At a BAC between 60 and 100 mg/dL euphoria is more marked, often causing excitement with partial or complete loss of inhibitions and in some individuals judgment and control are seriously impaired. When the BAC is between 100 and 150 mg/dL, which are concentrations seldom reached during moderate social drinking, psychomotor performance deteriorates markedly and poor articulation and speech impediment is obvious. Between 150 and 200 mg/dL ataxia is pronounced and drowsiness and confusion are evident in most people. The relationship between BAC and clinical impairment is well documented in drunk drivers who often reach very high BACs of 350 mg/dL or more, but most of these individuals are obviously chronic alcoholics.[213,214] In two recent studies of forensic autopsies, the average BAC when death was attributed to acute alcohol poisoning was 360 mg/dL.[215,216]

It is important to note that the impairment effects of alcohol depend to a great extent on the dose and the speed of drinking and whether the person starts from zero BAC or not.[217,218] The person's age and experience with alcohol are important owing to the development of central nervous system tolerance.[219] People who are capable of functioning with a very high BAC, such as drunk drivers, e.g., 200 to 300 mg/dL, have probably been drinking continuously for several days or even weeks so that a chronic tolerance to alcohol has had time to develop. Drinking a large volume of neat spirits in a short time results in nausea, gross behavioral impairment, and marked drunkenness, and an inexperienced drinker runs the risk of losing consciousness and suffering acute alcohol poisoning. Drinking too much too fast is dangerous, and if gastric emptying is rapid, the BAC rises with such a velocity that a vomit reflex in the brain is triggered. This physiological response to acute alcohol ingestion has probably saved many lives.

Conducting controlled studies with people who drink to reach very high BAC are difficult to motivate for ethical reasons. One exceptional study was reported by Zink and Reinhardt,[220] who

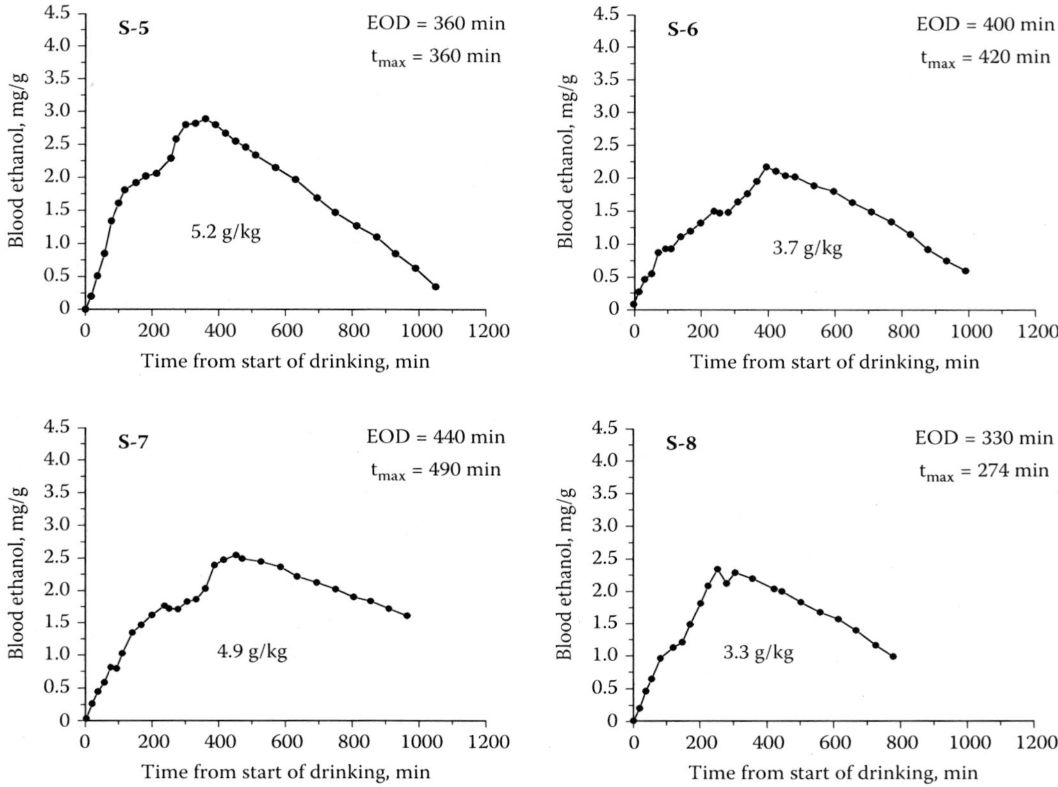

Figure 2.11 Concentration–time profiles of ethanol in four subjects (S-5 to S-8) who consumed very large quantities of alcohol (3.3 to 5.2 g/kg body weight) during a drinking time of 8 to 10 hours under controlled social conditions. EOD is time to end of drinking and t_{max} is time to reach the maximum BAC. Curves are redrawn from information published by Zink and Reinhardt.[220]

allowed healthy male volunteers to consume very large quantities of alcohol, either as beer or spirits or both, continuously for 8 to 10 h under social conditions. The BAC profiles were established unequivocally by frequent blood sampling from indwelling catheters every 15 to 20 min for up to 10 h. Some of the subjects reached abnormally high BAC of over 3.0 g/kg (300 mg/dL) and all developed a high degree of tolerance to the effects of alcohol. Figure 2.11 gives examples of the blood-concentration time profiles for four of the men who participated in this German study. Note that BAC is given in mass/mass (g/kg) as is customary in Germany and not weight/volume.

2.7 CLINICAL PHARMACOKINETICS OF ETHANOL

Clinical pharmacokinetics deals with the way that drugs and their metabolites are absorbed, distributed, and metabolized in the body and how these processes can be described in quantitative terms.[221–224]

2.7.1 Widmark Model

The clinical pharmacokinetics of ethanol have been investigated extensively since the 1930s thanks to the early availability of a reliable method of analysis in small volumes of blood.[209] Figure

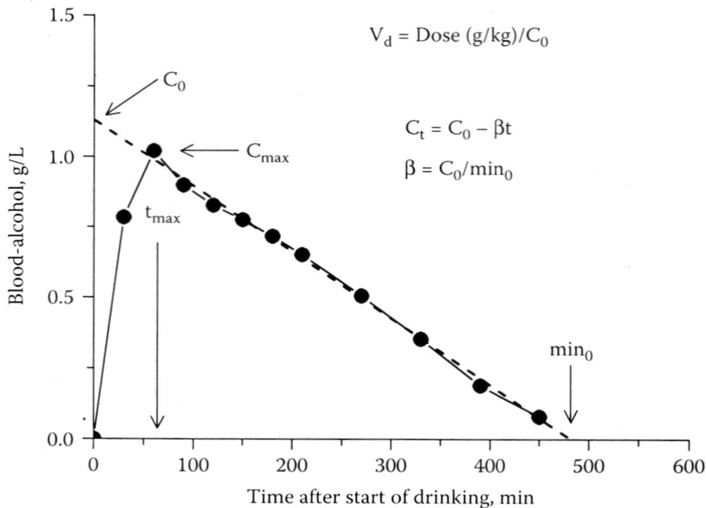

Figure 2.12 Blood-alcohol profile typically obtained after ingestion of a moderate dose of alcohol (0.68 g/kg) as neat whisky in 20 min after an overnight fast. Key pharmacokinetic parameters and how these are calculated are shown on the graph (see text for details).

2.12 shows a typical BAC–time profile after a healthy male subject drank 0.68 g/kg ethanol as neat whisky in the morning on an empty stomach. Before any pharmacokinetic evaluation of this curve is attempted, the data points on the post-absorptive phase should be carefully inspected and shown to fit well on a straight line. One indication of this is a high correlation coefficient ($r > 0.98$) for the concentration–time data points. The rectilinear declining phase is then extrapolated back to the ordinate (y-axis) or the time of starting to drink to give the C_0 parameter. This represents the concentration of ethanol in blood if the entire dose were absorbed and distributed in all body fluids and tissues without any metabolism occurring. The ratio of dose (g/kg) to C_0 (g/L) gives the ratio of body alcohol concentration to BAC and is known as the apparent volume of distribution, denoted "r" by Widmark or more recently V_d in units of L/kg. Inspection of this parameter allows a check on the validity of the experiment and the kinetic analysis because V_d can only take certain values. The value expected corresponds to the ratio of water in the whole body (60%) to the water content of the blood (80%) and thus a ratio of 0.6 to 0.7 L/kg for a healthy male and 0.5 to 0.6 L/kg for a female.[209]

Alcohol can also be administered intravenously, which is sometimes desirable in research and clinical investigations to avoid problems with variable gastric emptying and to avoid first-pass metabolism occurring in the stomach or the liver or both organs.[224] In the example shown in Figure 2.13, the test subject received 0.80 g/kg as a 10% w/v solution in saline at a constant rate for 40 min. The peak BAC now coincides with the end of the infusion period and this is followed by a diffusion plunge, during which time ethanol equilibrates between the well-perfused central blood compartment and poorly perfused resting skeletal muscle tissues. At about 90 min post-infusion, the BAC starts to decrease at a constant rate per unit time in accordance with zero-order kinetics and the slope of this rectilinear disappearance phase is commonly referred to as the alcohol burn-off rate or β-slope. However, specialist textbooks in pharmacokinetics refer to the zero-order elimination slope as k_0 instead of β.[224] When the blood concentration decreases below about 10 mg/dL or after about 450 min post-dosing in Figure 2.13, the linear declining phase becomes curvilinear for the remainder of time alcohol is still measurable in the blood.[223,224] The elimination of alcohol now follows first-order kinetics and the rate constant is denoted k_1. Some studies showed that the half-life of this terminal phase was about 15 min.[225]

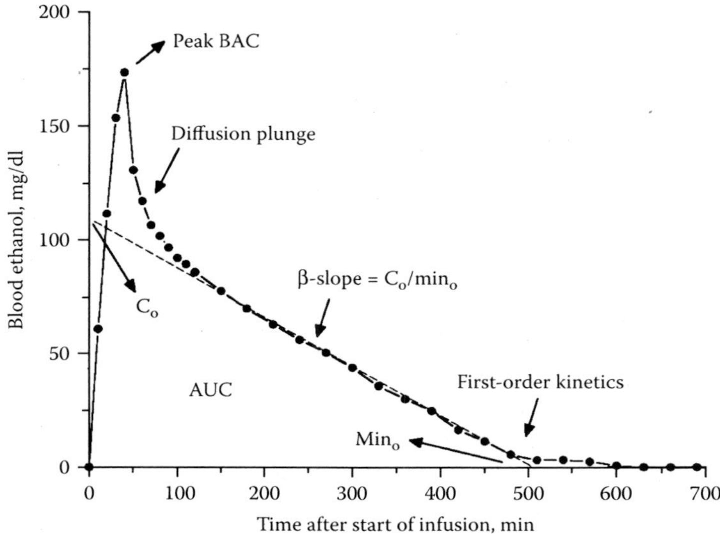

Figure 2.13 Concentration–time profile of ethanol in venous blood after one subject received 0.8 g ethanol per kg body weight by constant rate intravenous infusion over 40 min. The pharmacokinetic parameters are defined on this graph (see text for details).

The first person to make a comprehensive mathematical analysis of BAC profiles was Erik M.P. Widmark and details of his life and work have been published.[226] Widmark introduced the following equation to represent the elimination kinetics of alcohol from blood in the post-absorptive phase.

$$C_t = C_0 - \beta t \tag{1}$$

where C_t = blood alcohol concentration at some time t on the post-absorptive part of the curve, C_0 = blood alcohol concentration extrapolated to the time of starting to drink, β = rate of elimination of alcohol from blood, and t = time in minutes.

The rate of elimination of alcohol from the blood in moderate drinkers falls within the range 10 to 20 mg/dL/h with a mean value of about 15 mg/dL/h.[197,209,225] Higher values are seen in drinking drivers (mean 19 mg/dL/h)[227] and in alcoholics undergoing detoxification (mean 22 mg/dL/h).[228–231] The faster burn-off rates seen in heavy drinkers is probably one consequence of enzyme induction, which boosts the activity of the microsomal P4502E1 system during prolonged exposure to high concentrations of ethanol.[196,230,232] The P4502EI enzymes have a higher K_m (60 to 80 mg/dL) compared with ADH (2 to 5 mg/dL) and the slope of the elimination phase tends to be steeper when starting from a higher initial BAC, such as in alcoholics compared with moderate social drinkers.[230] In a controlled study with alcoholics undergoing detoxification, the mean β-slope was 22 mg/dL/h with a range from 13 to 36 mg/dL/h.[228] Liver disorders such as alcoholic hepatitis and cirrhosis did not seem to influence the rate of disposal of alcohol in these individuals.[228] When recently drinking alcoholics with high rates of alcohol elimination from blood were allowed to sober up for a few days and dosed again with a moderate amount of alcohol, the elimination rate was now in the range expected for moderate drinkers, namely, 15 mg/dL/h.[230]

The rate of elimination of alcohol from the blood was not much influenced by the time of day when 0.75 g/kg was administered at 9 A.M., 3 P.M., 9 P.M., and 3 A.M., according to an investigation into chrono-pharmacokinetics of ethanol.[233] However, gastric emptying seems to occur faster in the morning as reflected in a 32% higher peak BAC and an earlier time of its occurrence when ethanol (1.1 g/kg body weight) was consumed between 7.15 and 7.45 A.M., compared with the same time in the evening.[234] Smoking cigarettes slows gastric emptying and as a consequence delays the absorption of a moderate dose (0.50 g/kg) of ethanol resulting in a lower peak BAC in smokers.[235]

By extrapolating the rectilinear elimination phase back to the time of starting to drink, one obtains the y-intercept (C_0), which corresponds to the theoretical BAC expected if the entire dose was absorbed and distributed without any metabolism occurring (Figure 2.13). The empirically determined value of C_0 will always be greater than the ratio of dose/body weight because whole blood is 80% w/w water compared with the body, which is 60% w/w on average for men and 50% for women. The apparent volume of distribution (V_d) of alcohol is given by the ratio of dose (g/kg) divided by C_0 and in clinical pharmacology textbooks this is referred to as V_d with units of L/kg.[221–223] However, because BAC in Widmark's studies was reported in units of mg/g or g/kg, dividing the dose by C_0 gives a ratio without any dimensions. This needs to be considered whenever BAC is reported as weight/volume units (e.g., g/L) as is more usual today. The density of whole blood is 1.055 g/mL, which means there is an expected difference of 5.5% compared with values of V_d reported by Widmark.[209]

Values of the distribution factor "r" differ between individuals depending on age and body composition particularly the proportion of fat to lean body mass.[236] Obviously, the value of "r" will also depend on whether whole blood or plasma specimens were analyzed and used to plot the concentration–time profile when back-extrapolation is done to determine C_0. As shown in Figure 2.2, the plasma–alcohol curves run on a higher level compared with whole blood–alcohol curves because of the differences in water content as discussed earlier. This means that C_0 is higher for plasma curves compared with whole-blood curves.[47]

According to Widmark's second equation, the relationship between alcohol in the body and alcohol in the blood at equilibrium can be represented by the following equations:

$$A/(p \times r) = C_0 \tag{2}$$

$$A = C_0 \times (p \times r) \tag{3}$$

where A = amount of alcohol in grams absorbed and distributed in all body fluids, p = body weight of the person in kg, r = Widmark's "r" factor, and C_0 = y-intercept (Figure 2.13).

These equations make it is easy to calculate the amount of alcohol in the body from the concentration determined in a sample of blood provided that the value of "r" is known and that absorption and distribution of ethanol were complete at the time of sampling blood. However, in reality 100% absorption of the dose is only achieved when ethanol is given by intravenous infusion. Thus, the above equation will tend to overestimate the person's true BAC because part of an orally administered dose might be cleared by first-pass metabolism occurring either in the stomach or liver or in both places. Although the above equation has been much used in forensic alcohol calculations, it should not be applied to other drugs and narcotics and especially not in post-mortem toxicology, owing to problems with variation in drug concentrations in different sampling sites (see Section 2.3).

In the fasting state, the factor "r" will depend on age, gender, and body composition and Widmark reported mean values of 0.68 for 20 men (range 0.51 to 0.85) and 0.55 for 10 women (range 0.49 to 0.76).[209] However, in many later studies, which included more volunteer subjects, it was found that average values of "r" were closer to 0.70 L/kg for men and 0.60 L/kg for women with 95% confidence limits of about ±20%.[237]

The two separate Widmark equations for β and "r" can be easily combined by eliminating C_0 to give the following equation:

$$A = pr(C_t + \beta t) \tag{4}$$

The above equation is useful to estimate the total amount of alcohol absorbed from the gastrointestinal tract since the beginning of drinking or by rearrangement, the BAC (C_t) expected after intake of a known amount of alcohol.

$$C_t = (A/pr) - \beta t \tag{5}$$

When calculating BAC from the dose administered, or vice versa, it is necessary to assume that systemic availability is 100% and that absorption and distribution of alcohol into total body water is complete at the time of sampling blood. Furthermore, individual variations in β and "*r*" introduce uncertainty in the calculated dose (*A*) or BAC (C_t) when average values are applied to random subjects from the population. Various modifications or improvements have been suggested, such as by using estimates of total body water, lean body mass, or nomograms based on body mass index.[236–238] The individual variation has been estimated to ±20% for 95% confidence limits in tests involving more than 100 subjects who drank alcohol on an empty stomach.[237] However, in the entire population of drinking drivers, these limits can be expected to be much wider.

2.7.2 Michaelis–Menten Model

Because the class I ADH enzymes have a low k_m (2 to 5 mg/dL) for ethanol they become saturated with substrate after one to two drinks.[221,222] The rate of disappearance of ethanol from blood therefore follows zero-order kinetics over a large segment of the post-absorptive elimination phase (Figure 2.13).[236,237] When the BAC decreases below about 10 mg/dL, the ADH enzymes are no longer saturated and the curve changes to a curvilinear disappearance phase (first-order kinetics).[221–223,239] However, these low BACs are not very relevant when forensic science aspects of ethanol are concerned.

It was first demonstrated by Lundquist and Wolthers[240] that the entire post-absorptive elimination phase (zero-order and first-order stages) might be rationalized by an alternative pharmacokinetic model, namely, that of saturation kinetics.[223,239] Thanks to the availability of a highly sensitive ADH method of analysis, much lower blood-alcohol concentration (<10 mg/dL) could be reliably determined. These investigators from Denmark fitted their BAC–time data to the Michaelis–Menten (M-M) equation and from its integrated form they arrived at the parameters V_{max} and k_m that described the enzymatic reaction.[240] This application of nonlinear saturation kinetics of ethanol was thereafter strongly advocated by many specialists in pharmacokinetics, among others, Wagner, Wilkinson, and their colleagues[221,222,241–243] and values of 22 mg/dL/h and 5 mg/dL were suggested for V_{max} and k_m, respectively.[222] Although the use of M-M kinetics has found some support among forensic scientists,[243,244] others have not considered this necessary for actual casework because so many other variable factors and uncertainties influence the absorption, distribution, and elimination pattern of ethanol.[244] Moreover, the mathematical concepts needed to understand and apply M-M kinetics are much more challenging than those necessary to derive the Widmark equation. Explaining the scientific principles of pharmacokinetic multicompartment models and nonlinear kinetics to a judge and jury is a daunting task. Moreover, the idea of multiple enzyme systems being involved in the metabolism of ethanol including racial and ethnic differences and polymorphism might overly complicate the situation.[245,246] The contribution of various isozymes of ADH as well as the higher k_m microsomal CYP2E1 pathway for disposal of ethanol after chronic ingestion (metabolic tolerance) explains the faster rate of clearance from blood in binge drinkers.[201,204] Furthermore, the contribution from several enzyme systems is not strictly compatible with the conditions assumed to apply for application of the classical M-M equation.

2.7.3 First-Pass Metabolism and Gastric ADH

About 20 years ago, considerable attention was drawn to the possibility that metabolism of ethanol occurred also pre-systemically in the gastric mucosa.[247–252] Accordingly, part of the dose of ethanol ingested was broken down before reaching the systemic circulation and this was thought to involve an enzymatic oxidation process taking place in the gastric mucosa. Pre-systemic removal of an administered drug is important to understand and has implications for therapeutic efficacy. However, the liver, which contains most of the metabolizing enzymes, was considered the primary site for first-pass metabolism (FPM) of ethanol to occur. The magnitude of FPM tended to exhibit large intersubject variations and was dose dependent and strikingly influenced by speed of gastric emptying as well as fed–fasted state.[252] FPM was more pronounced after very small doses of ethanol (0.15 to 0.3 g/kg) when this was consumed about 1 h after eating a meal.[248,249] Food tends to delay stomach emptying thus allowing more time for ethanol metabolism by enzymes located in the gastric mucosa.

Distinguishing between pre-systemic oxidation of ethanol occurring in the stomach as opposed to the liver proved a very difficult task and much debate arose about the significance of gastric ADH compared with hepatic ADH as the site for FPM.[136,250–252] Some workers were adamant about the importance of gastric FPM of ethanol, whereas others challenged its importance in the overall disposition of ethanol in the body.[250–253] The argument against rested in part on the fact that gastric ADH represented only a small fraction of the total amount of hepatic ADH and also the complex nature of M-M kinetics especially at the low BAC reached after small doses were ingested.[250] The main advocates of gastric ADH and its role in FPM of ethanol had failed to consider the critical importance of stomach emptying and effective clearance of ethanol at low substrate concentrations.[250–252] Factors that influence the rate of absorption of alcohol (food, drugs, type of beverage, posture, time of day) also influence the concentration of alcohol entering the portal venous blood and the degree of saturation of ethanol-metabolizing enzymes in the liver.[245–252]

Interest in FPM and gastric ADH escalated considerably when several publications appeared showing that commonly used medications, such as aspirin and H_2-receptor antagonists (cimetidine and ranitidine), when combined with small amounts of alcohol, resulted in higher C_{max} and area under the curve (AUC) compared with no-drug or placebo control treatment.[253–256] These drugs were shown to block activity of gastric ADH under *in vitro* conditions (e.g., in biopsies from the stomach) but the effects *in vivo* were less clear-cut.[257] Proponents of the role of gastric ADH in FPM of alcohol argued that this represented a protective barrier against some of the toxic effects of ethanol that cause organ and tissue damage.[258] If this barrier were weakened or removed by various common medications, this would result in greater exposure of inner organs to ethanol and a higher than expected C_{max} and, accordingly, a more pronounced impairment of body functions.[253–256,258] Because the activity of gastric ADH was lower in women, Asian populations, and also in alcoholics, these groups of people were thought to be more vulnerable to the toxic effects of moderate drinking.[249,259] This conclusion proved too hasty because a flood of articles appeared challenging the notion of drug-induced effects on gastric FPM of ethanol.[260–265] Indeed, these newer studies incorporated an improved experimental design with many more volunteer subjects and a wider range of alcohol doses, times of day, and intake with or without food.[260–265]

What did emerge from this new wave of interest in the clinical pharmacokinetics of ethanol was strong and convincing information for large inter- and intra-individual variations in the pharmacokinetic profiles of ethanol especially when small doses (0.15 to 0.3 g/kg) were ingested after a meal.[266–268] The biological variation in blood-alcohol profiles was particularly evident when small doses of alcohol were ingested together with or after a meal.[269,270]

2.7.4 Food and Pharmacokinetics of Ethanol

The presence of food in the stomach before drinking retards the absorption of ethanol from the gut by delaying gastric emptying.[271–273] After a meal, the C_{max} was considerably lower and the impairment effects of ethanol diminished compared with drinking the same dose on an empty stomach.[273] The reduced bioavailability of ethanol did not seem to depend on the composition of the meal in terms of its protein, fat, or carbohydrate content.[274–276] The altered bioavailability of ethanol with food in the stomach before drinking obviously has consequences when the apparent volume of distribution of ethanol and other kinetic parameters are evaluated, such as AUC. The dramatic lowering of C_0 in the fed state means that the ratio dose/C_0 might increase to reach an impossible nonphysiological result (>1.0) compared with the value expected of 0.7 in men and 0.6

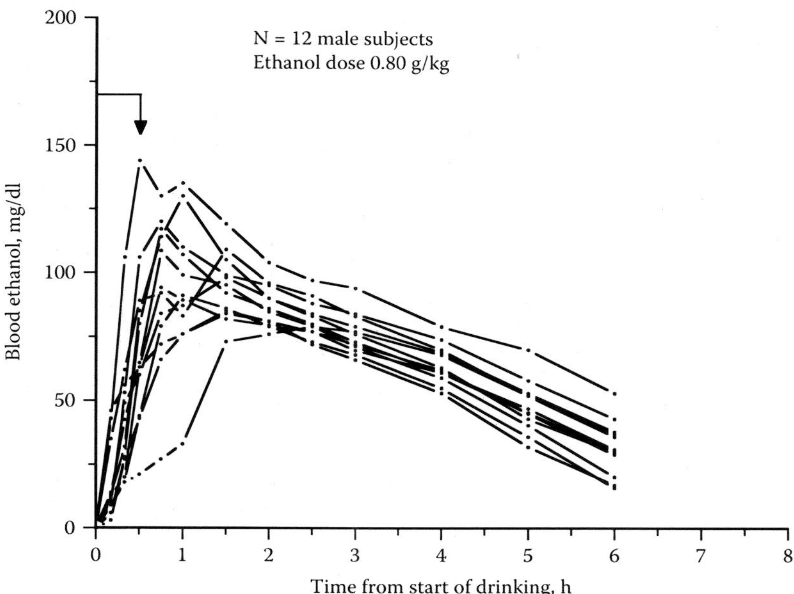

Figure 2.14 Individual concentration–time profile of ethanol in venous blood for 12 healthy men who drank 0.80 g ethanol per kg body weight in 30 min after an overnight fast.

in women. The notion of a "loss of ethanol" when drinking occurred after a meal was observed by Widmark,[109] and he proposed that the mechanism might involve chemical reaction of ethanol with constituents of the food. However, a more modern explanation for this food-induced lowering in the bioavailability of ethanol is presystemic oxidation by gastric and/or hepatic ADH.[274,275]

Figure 2.14 shows individual BAC curves after 12 male subjects drank 0.8 g/kg in the morning after an overnight fast.[268] Both C_{max} and t_{max} varied widely between individuals. The time to reach the C_{max} ranged from 0 to 150 min after the end of drinking. Those subjects with very slow rates of absorption might have experienced a pyloric spasm so that the absorption took place through the stomach as opposed to the duodenum and jejunum where rapid absorption occurs.

The impact of food and body composition on BAC–time profiles for various drinking scenarios was recently the subject of a comprehensive review by Kalant.[277] Ingestion of food immediately before or together with alcohol resulted in a lower C_{max} compared with drinking the same dose on an empty stomach. This lowering effect does not seem to be related to the composition of the meal in terms of macronutrients (protein, fat, or carbohydrate), but instead the size of the meal is seemingly more important.[274] The delayed gastric emptying meant a slower delivery of alcohol into the small intestine, the site of rapid absorption into the portal blood, and therefore a longer exposure to gastric mucosal ADH thus enhancing the changes of gastric FPM of alcohol. Moreover, the role of FPM, whether gastric or hepatic, is small and highly variable and much depends on the dose of alcohol administered. After very small doses (<0.3 g/kg) FPM is more pronounced compared with moderate social drinking when BACs in the forensic relevant range (50 to 80 mg/dL) are reached.[277]

The BAC traces shown in Figure 2.15 demonstrate a marked effect of eating a meal before drinking. Both the C_{max} and also the area under the concentration–time profile are considerably lowered compared with drinking the same dose on an empty stomach.[189] Here the subjects drank 0.8 g ethanol per kg either after an overnight (10 h) fast or immediately after eating a standardized breakfast. Drinking in the fed state also results in a faster metabolism of ethanol as reflected in the shorter time of 1 to 2 h needed to reach zero BAC.[189] This accelerating effect of a meal on the elimination rate of alcohol from blood was confirmed even when the alcohol was given intravenously.[278,279] The mechanism of food-induced acceleration of ethanol metabolism might be related

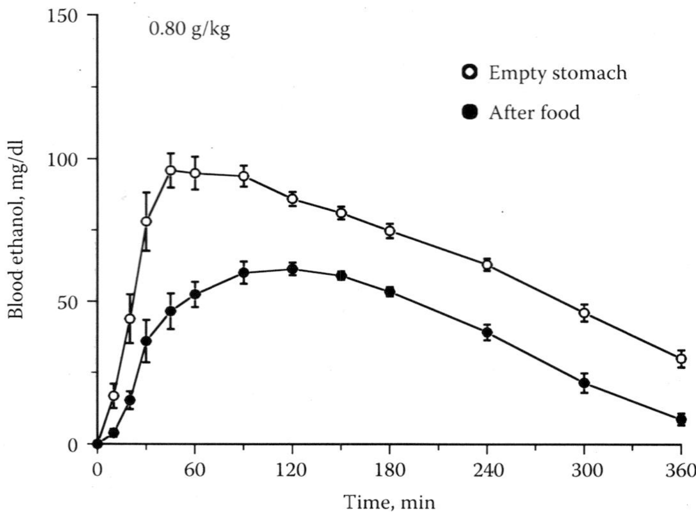

Figure 2.15 Mean blood-alcohol profiles after eight healthy men drank alcohol (0.80 g/kg in 30 min) either on an empty stomach or immediately after eating breakfast.

to an enhanced hepatic enzyme activity in a well-nourished organism. Furthermore, eating a meal is known to increase liver blood flow, which also facilitates a more rapid clearance of ethanol from the bloodstream at low ethanol concentrations.[275,276,278]

2.8 CONCLUDING REMARKS

Harmful and hazardous consumption of alcohol is increasing worldwide owing to, among other things, the lowering of excise taxation in some countries and longer opening times for bars and restaurants in others.[6] History has shown that deaths attributable to overconsumption of alcohol generally increase as a function of total consumption of alcohol in a population.[5,6] Chronic heavy drinking eventually leads to a wide spectrum of disease states and health disorders, particularly liver diseases including hepatitis, cirrhosis, and death.[280] Acute inflammation of the pancreas is another consequence of heavy drinking, not to mention a host of unintentional injuries associated with drunkenness that require emergency medical treatment.[280] Drinking five drinks or more on one occasion is considered binge drinking, and might result in acute intoxication and drunkenness and trigger different kinds of reckless behavior, such as unsafe sex, drunk driving, aggressiveness, and family violence.[7–10] There is not the slightest doubt that testing for alcohol will remain the most frequently requested procedure from toxicology laboratories for a long time to come.

Few substances can be measured with such a high degree of accuracy, precision, and selectivity as the concentration of ethanol in a person's blood.[17] The method of headspace gas chromatography is the gold standard and can be used to analyze a wide range of volatile substances in body fluids besides ethanol.[281,282] Making duplicate determinations is an effective safeguard against various mishaps that might occur during sample preparation and analysis. The aliquots of blood should whenever possible be taken from two separate Vacutainer tubes, which will help to minimize the risk of sample mix-up during processing. To enhance the analytical selectivity, the duplicates should be run on different chromatographic systems, thus yielding different retention times for the substances of interest.[17,23–25] Alternatively, an independent analytical method such as enzymatic or chemical oxidation or mass spectrometry might be used in parallel with HS-GC to enhance specificity for ethanol.[131–133]

More attention should be given to pre-analytical factors that might influence precision and accuracy of the results of ethanol determination. This includes preparation of the subject, the way the blood sample was drawn including the sampling site and the skin disinfection used, as well as the kind of Vacutainer tubes and the preservatives they might contain. Things like the ambient temperature conditions, the mode and duration of transportation of blood specimens to the laboratory, and the condition of the blood sample on arrival (volume submitted and whether coagulated or not) are of concern. The situation is compounded when blood samples are taken from victims of traffic accidents who might require emergency hospital treatment including the administration of drugs or intravenous fluids to counteract shock. It is important to remember that the result of a chemical or biochemical test is only as good as the sample received for analysis. When the analytical result is close to the legal limit for driving, small errors that creep into the analytical procedures might make the difference between punishment or acquittal in borderline cases. Thus, pre-analytical factors are just as important to control and document as analytical factors.

In forensic casework when the BAC exceeds 0.2 g/L (20 mg/100 mL), the elimination of alcohol from blood occurs at a constant rate per unit time (zero-order kinetics) as was first proposed by Widmark.[190,209] From knowledge about the distribution volume of ethanol and the measured BAC, it is easy to calculate the amount of alcohol absorbed and distributed in the body at the time of blood sampling. Also if the time of starting to drink is known, then the total amount of alcohol consumed can be calculated by adding on the amount of alcohol eliminated through metabolism.[283] A good rule of thumb is that the rate of ethanol elimination from the whole body for a moderate drinker is approximately 0.1 g/kg body weight per h and this rate seems to be independent of gender.[284,285] Making a back-extrapolation of a person's BAC from the time of sampling to the time of driving is a dubious practice.[286] Without knowledge of the prior drinking scenario, including the timing and quantity of alcohol in the last drink, it is difficult to predict with certainty when C_{max} occurs. Consideration should also be given to inter- and intra-individual variations in the pharmacokinetic parameters of ethanol, especially the rate of alcohol disappearance from the bloodstream.[186,187]

Warnings about adverse drug–alcohol interactions and the risk of experiencing an unexpectedly high BAC after moderate drinking owing to concomitant use of medication such as aspirin and antacids (ranitidine and cimetidine) have been much exaggerated.[200] The magnitude of such an effect measured in percent was greatest after drinking very small doses of alcohol 1 h after subjects ate a fatty meal.[248,249] However, it is worth reiterating that small absolute increases in C_{max} between drug-treated and control group (e.g., 5 mg/dL) yield large percentage differences when the denominator of the ratio is low (e.g., 15 mg/dL); this corresponds to a 33% change. When subjects drank small doses of ethanol (0.15 g/kg) repetitively (four times) over a few hours, which is more like social drinking conditions, after medication with Zantac (ranitidine), both the C_{max} and AUC were higher compared with a placebo control treatment.[287] However, this observation still needs to be confirmed by other investigators. The authors of the study, which was published in a reputable scientific journal, concluded that social drinkers who might be medicated with ranitidine should be warned of the risk of developing unexpected functional impairment after drinking amounts of alcohol they might normally consider safe. Unfortunately, tests of functional impairment were not included in the study design so the warning is speculative and unfounded.[287]

The inter- and intra-individual differences in pharmacokinetics of alcohol were recently reviewed with major focus on Michaelis–Menten and multicompartment models.[288] Some investigators advocate use of more complex three-compartment models to explore the disposition and fate of alcohol in the body and to explain these processes in quantitative terms.[289–292] Within the forensically relevant range of BAC (20 to 300 mg/dL), the zero-order kinetics of ethanol first introduced by Widmark roughly 70 years ago still remains valid. Age and gender differences in the pharmacokinetics of ethanol are largely accounted for by differences in total body water (TBW) and therefore the distribution volume of ethanol (Widmark's r factor or V_d).[277] The notion of a significant gastric first-pass metabolism, including racial- and gender-related differences, has been much exaggerated and is only relevant after tiny doses of alcohol (<0.3 g/kg) are ingested in the fed state.[250–252] After moderate doses of alcohol have been consumed on an empty stomach, the bioavailability of ethanol is close to 100% and under these conditions TBW can be calculated by ethanol dilution, which speaks against any significant first-pass metabolism.[47,190,277] Ethanol's special physicochemical and physiological properties means that a person's BAC can be interpreted in relation to the amount of alcohol ingested.[283] Such calculations are not feasible for other drugs of abuse.

This overview has made it abundantly clear that ethanol is no simple drug, neither in terms of its pharmacokinetics nor its pharmacodynamics and mechanism of action on the brain. The costs to society for treatment and rehabilitation of people with alcohol problems are astronomical.[293] The entire field of biomedical alcohol research with the main focus on alcohol-related pathology was recently published in three volumes.[280] This comprehensive handbook is the latest rendition of basic and clinical aspects of the alcohol research literature and is well worth referencing.[280]

REFERENCES

1. Klatsky, A.L., Drink to your health. *Sci. Am.,* 288, 62–69, 2003.
2. Murphy, J.G., McDevitt-Murphy, M.E., and Barnett, N.P., Drink and be merry? Gender, life satisfaction and alcohol consumption among college students. *Psychol. Addict. Behav.,* 19, 184, 2005.
3. Gibbons, B., Alcohol — the legal drug. *Natl. Geogr.,* 181, 3, 1992.
4. Mukamal, K.J., Conigrave, K.M., Mittleman M.A., Camargo, C.A., Jr., Stampfer, M.J., Willett, W.C., and Rimm, E.B., Roles of drinking pattern and type of alcohol consumed in coronary heart disease in men. *N. Engl. J. Med.,* 348, 109, 2003.
5. Saitz, R., Unhealthy alcohol use. *N. Engl. J. Med.,* 352, 596, 2005.
6. Room, R., Babor, T., and Rehm, J., Alcohol and public health. *Lancet,* 365, 519, 2005.
7. Cherpitel, C.J., Alcohol and casualties — a comparison of emergency room and coroner data. *Alc. Alcohol.,* 29, 211, 1994.
8. Cherpitel, C.J., Injury and the role of alcohol; county-wide emergency room data. *Alcohol Clin. Exp. Res.,* 18, 679, 1994.
9. Stewart, S.H., Alcohol abuse in individuals exposed to trauma; a critical review. *Psychol. Bull.,* 120, 83, 1996.
10. VonMoreau, K.B., Mueller, P., Drirsch, D., Osswald, B., and Seitz, H.K., Alcohol and trauma. *Alcohol Clin. Exp. Res.,* 16, 141, 1992.
11. Driscoll, T.R., Harrison, J.A., and Steenkamp, M., Review of the role of alcohol in drowning associated with recreational aquatic activity. *Inj. Prev.,* 10, 107, 2004.
12. Cummings, P. and Quan, L., Trends in unintentional drowning: the role of alcohol and medical care. *J. Am. Med. Assoc.,* 281, 2198, 1999.
13. Chen, L.H., Baker, S.P., and Li, G., Drinking history and risk of fatal injury: comparison among specific injury causes. *Accid. Anal. Prev.,* 37, 245, 2005.
14. Jones, A.W., Forensic sciences; determination of alcohol in body fluids, in *Encyclopedia of Analytical Sciences*, Academic Press, New York, 1995, 1585.
15. Wright, J.W., Alcohol and the laboratory in the United Kingdom. *Ann. Clin. Biochem.,* 28, 212, 1991.
16. Garriott, J., *Medicolegal Aspects of Alcohol,* Lawyers & Judges, Tucson, 2004.
17. Jones, A.W., Measuring alcohol in blood and breath for forensic purposes — a historical review. *Forensic Sci. Rev.,* 8, 13, 1996.
18. Quinlan, K.P., Brewer, R.D., Sigel, P., Sleet, D.A., et al., Alcohol-impaired driving among US adults 1993–2002. *Am. J. Prev. Med.,* 28, 346, 2005.
19. Ferrara, S.D., Low blood-alcohol concentrations and driving impairment. A review of experimental studies and international legislation. *Int. J. Legal Med.,* 106, 169, 1994
20. Jones, A.W., Enforcement of drink-driving laws by use of "per se" legal alcohol limits; blood and/or breath alcohol concentration as evidence of impairment. *Alc. Drugs Driving,* 4, 99, 1988.
21. Ziporyn, T., Definition of impairment essential for prosecuting drunken drivers. *J. Am. Med. Assoc.,* 253, 3509, 1985.
22. Dubowski, K.M. and Caplan, Y., Alcohol testing in the workplace, in *Medicolegal Aspects of Alcohol,* Garriott, J., Ed., Lawyers & Judges, Tucson, 1996, 439.

23. Dubowski, K.M., Recent developments in alcohol analysis. *Alc. Drugs Driving,* 2, 13, 1986.
24. Dubowski, K.M., Alcohol determination in the clinical laboratory. *Am. J. Clin. Pathol.,* 74, 747, 1980.
25. Jones, A.W. and Schuberth, J.O., Computer-aided headspace gas chromatography applied to blood-alcohol analysis; importance of on-line process control. *J. Forensic Sci.,* 34, 1116, 1989.
26. Kelly, D.F., Alcohol and head injury, an issue revisited. *J. Neurotrauma,* 12, 883, 1995.
27. Rutherford, W.H., Diagnosis of alcohol ingestion in mild head trauma. *Lancet,* 1, 1021, 1977.
28. Quaghebeur, G. and Richards, P., Comatose patients smelling of alcohol. *Br. Med. J.,* 299, 410, 1989.
29. Pappas, S.C. and Silverman, M., Treatment of methanol poisoning with ethanol and hemodialysis. *CMA J.,* 15, 1391, 1982.
30. Jacobsen, O. and McMartin, K.E., Methanol and ethylene glycol poisoning; mechanisms of toxicity, clinical course, diagnosis and treatment. *Med. Toxicol.,* 1, 309, 1986.
31. Walder, A.D. and Tyler, C.K.G., Ethylene glycol antifreeze poisoning. *Anaesthesia,* 49, 964, 1994.
32. Egarbane, B., Borron, S.W., and Baud, F.J., Current recommendations for treatment of severe toxic alcohol poisonings, *Intensive Care Med.,* 31, 189, 2005.
33. Flanagan, R.J. and Jones, A.L., *Antidotes,* Taylor & Francis, London, 2001.
34. Youssef, G.M. and Hirsch, D.J., Validation of a method to predict required dialysis time for cases of methanol and ethylene glycol poisoning. *Am. J. Kidney Dis.,* 46, 509, 2005.
35. Brent, J., McMartin, K., Phillips, S., Aaron, C., and Kulig, K., Fomepizole for the treatment of methanol poisoning. *N. Engl. J. Med.,* 344, 424, 2001.
36. Eder, A.F., McGrath, C.M., Dowdy, Y.G., Tomaszewski, J.E., Rosenberg, F.M., et al., Ethylene glycol poisoning: toxicokinetics and analytical factors affecting laboratory diagnosis. *Clin. Chem.,* 44, 168, 1998.
37. Flanagan, R.J., SI units — common sense not dogma is needed. *Br. J. Clin. Pharmacol.,* 39, 589, 1995.
38. Flanagan, R.J., Guidelines for the interpretation of analytical toxicology results and unit of measurement conversion factors. *Ann. Clin. Biochem.,* 35, 261, 1998.
39. Lentner, C., *Geigy Scientific Tables,* Vol., *Units of Measurement, Body Fluids, Composition of the Body, Nutrition,* CIBA-GEIGY, Basel, 1981.
40. Burtis, C.A. and Ashwood, E.R., Eds., *Tietz Fundamentals of Clinical Chemistry,* 5th ed., W.B. Saunders, Philadelphia, 2001.
41. Iffland, R., West, A., Bilzer, N., and Schuff, A., Zur Zuverlässigkeit der Blutalkoholbestimmung. Das Verteilungsverhältnis des Wassers zwischen Serum und Vollblut. *Rechtsmedizin,* 9, 123, 1999.
42. De Jong, G.M.Th., Huizenga, J.R., and Gips, C.H., Evaluation of gravimetric assays of the H2= concentration in human serum and urine. *Clin. Chim. Acta,* 163, 153, 1987.
43. Winek, C.L. and Carfagna, M., Comparison of plasma, serum, and whole blood ethanol concentrations. *J. Anal. Toxicol.,* 11, 267, 1987.
44. Rainey, P.M., Relation between serum and whole blood ethanol concentrations. *Clin. Chem.,* 39, 2288, 1993.
45. Frajola, W.J., Blood alcohol testing in the clinical laboratory: problems and suggested remedies. *Clin. Chem.,* 39, 377, 1993.
46. Harmenina, D.M., Ed., *Clinical Hematology and Fundamentals of Hemostasis,* 4th ed., F.A. Davis, Philadelphia, 2002.
47. Jones, A.W., Hahn, R.G., and Stalberg, H.P., Pharmacokinetics of ethanol in plasma and whole blood: estimation of total body water by the dilution principle. *Eur. J. Clin. Pharmacol.,* 42, 445, 1992.
48. Jones, A.W., Hahn, R.G., and Stalberg, H.P., Distribution of ethanol and water between plasma and whole blood; inter- and intra-individual variations after administration of ethanol by intravenous infusion. *Scand. J. Clin. Lab. Invest.,* 50, 775, 1990.
49. Charlebois, R.C., Corbett, M.R., and Wigmore, J.G., Comparison of ethanol concentrations in blood, serum, and blood cells for forensic applications. *J. Anal. Toxicol.,* 20, 171, 1996.
50. Fung, W.K., Chan, K.L., Mok, V.K.K., Lee, C.W., and Choi, V.M.F., The statistical variability of blood alcohol concentration measurements in drink-driving cases. *Forensic Sci. Int.,* 110, 207, 2000.
51. Jones, A.W., Dealing with uncertainty in chemical measurements. *IACT Newsl.,* 14, 6, 2003.
52. Walls, H.J. and Brownlie, A.R., *Drink, Drugs and Driving,* 2nd ed., Sweet & Maxwell, London, 1985.
53. Fraser, C.G., *Interpretation of Clinical Chemistry Laboratory Data.* Blackwell Scientific, Oxford, 1986.
54. Kratz, A., Ferraro, M.J., Sluss, P.M., and Lewandrowski, K.B., Laboratory reference values. *N. Engl. J. Med.,* 351, 1548, 2004.

55. Widmark, E.M.P., Eine Mikromethode zur Bestimmung von Äthylalkohol im Blut. *Biochem. Z.,* 131, 473, 1922.
56. Friedemann, T.E. and Dubowski, K.M., Chemical testing procedures for the determination of ethyl alcohol. *J. Am. Med. Assoc.,* 170, 47, 1959.
57. Lundquist, F., The determination of ethyl alcohol in blood and tissue, in *Methods of Biochemical Analysis,* Vol. VII, Glick, D., Ed., Interscience, New York, 1959, 217.
58. Smith, H.W., Methods for determining alcohol, in *Methods of Forensic Science*, Curry, A.S., Ed., Interscience, New York, 1965, 3.
59. Jain, N.C. and Cravey, R.H., Analysis of alcohol. I. A review of chemical and infrared methods. *J. Chromatogr. Sci.,* 10, 257, 1972.
60. Bonnichsen, R.K. and Wassén, A., Crystalline alcohol dehydrogenase from horse liver. *Arch. Biochem.,* 18, 361, 1948.
61. Bonnichsen, R.K. and Theorell, H., An enzymatic method for the microdetermination of ethanol. *Scand. J. Clin. Lab. Invest.,* 3, 58, 1951.
62. Bücher, Th. and Redetzki, H., Eine spezifische photometrische Bestimmung von Äthylalkohol auf fermentivem Wege. *Klin. Wochnschr.,* 29, 615, 1951.
63. Redetzki, H. and Johannsmeier, K., Grundlagen und Ergebnisse der enzymatischen Äthylalkohol-bestimmung. *Arch. Toxikol.,* 16, 73, 1957.
64. Vasiliades, J., Pollock, J., and Robinson, A., Pitfalls of the alcohol dehydrogenase procedure for the emergency assay of alcohol: a case study of isopropanol overdose. *Clin. Chem.,* 24, 383, 1978.
65. Buijten, J.C., An automated ultra-micro distillation technique for determination of ethanol in blood and urine. *Blutalkohol,* 12, 393, 1975.
66. Kristoffersen, L., Skuterud, B., Larssen, B.R., Skurtveit, S., and Smith-Kielland, A., Fast quantitation of ethanol in whole blood specimens by the enzymatic alcohol dehydrogenase method. Optimization by experimental design. *J. Anal. Toxicol.,* 29, 66, 2005.
67. Whitehouse, L.W. and Paul, C.J., Micro-scale enzymatic determination of ethanol in plasma with a discrete analyzer, the ABA-100. *Clin. Chem.,* 25, 1399, 1979.
68. Hadjiioannou, T.P., Hadjiioannou, S.I., Avery, J., and Malmstedt, H.V., Automated enzymatic determination of ethanol in blood serum and urine with a miniature centrifugal analyzer. *Clin. Chem.,* 22, 802, 1976.
69. Siagle, K.M. and Ghosen, S.J., Immunoassays: tools for sensitive, specific, and accurate test results. *Lab. Med.,* 27, 177, 1996.
70. Caplan, Y. and Levine, B., The analysis of ethanol in serum, blood, and urine: a comparison of the TDx REA ethanol assay with gas chromatography. *J. Anal. Toxicol.,* 10, 49, 1986.
71. Urry, F.M., Kralik, M., Wozniak, E., Crockett, H., and Jennison, T.A., Application of the Technicon Chem 1 + chemistry analyzer to the Syva Emit ethyl alcohol assay in plasma and urine. *J. Anal. Toxicol.,* 17, 287, 1993.
72. Hannak, D. and Engel, C.H., Schnellbestimmung des Blutalkohols mit der ADH/REA methode: Methodenvergleich und Bewertung. *Blutalkohol,* 22, 371, 1985.
73. Alt, A. and Reinhardt, G., Die Genauigkeit der Blutalkoholbestimmung mit Head-Space GC, ADH und dem REA Ethanol Assay für das AXSYM System — ein Methodenvergleich. *Blutalkohol,* 33, 209, 1996.
74. Bland, J.M. and Altman, D.G., Statistical methods for assessing agreement between two methods of clinical measurement. *Lancet,* 1, 307, 1986.
75. Nine, J.S., Moraca, M., Virji, M.A., and Rao, K.N., Serum-ethanol determination: comparison of lactate and lactate dehydrogenase interference in three enzymatic assays. *J. Anal. Toxicol.,* 19, 192, 1995.
76. Sloop, G., Hall, M., Simmondon, G.T., and Robinson, C.A., False positive postmortem EMIT drugs of abuse due to lactate dehydrogenase and lactate in urine. *J. Anal. Toxicol.,* 19, 554, 1995.
77. Badcock, N.R. and O'Reilly, D.A., False positive EMIT-st ethanol screen with infant plasma. *Clin. Chem.,* 38, 434, 1992.
78. Cadman, W.J. and Johns, T., Application of the gas chromatography in the laboratory of criminalistics. *J. Forensic Sci.,* 5, 369, 1960.
79. Fox, J.F., Gas chromatographic analysis of alcohol and certain other volatiles in biological material for forensic purposes. *Proc. Soc. Exp. Biol. Med.,* 97, 236, 1958.

80. Chundela, B. and Janak, J., Quantitative determination of ethanol besides other volatile substances in blood and other body liquids by gas chromatography. *J. Forensic Med.,* 7, 153, 1960.
81. Parker, K.D., Fontan, C.R., Yee, J.L., and Kirk, P.L., Gas chromatographic determination of ethyl alcohol in blood for medicolegal purposes; separation of other volatiles from blood or aqueous solution. *Anal. Chem.,* 34, 1234, 1962.
82. Curry, A.S., Walker, G.W., and Simpson, G.S., Determination of alcohol in blood by gas chromatography. *Analyst,* 91, 742, 1966.
83. Machata, G., The advantages of automated blood alcohol determination by head space analysis. *Z. Rechtsmedizin.,* 75, 229, 1975.
84. Jain, N.C. and Cravey, R.H., Analysis of alcohol. II. A review of gas chromatographic methods. *J. Chromatogr. Sci.,* 10, 263, 1972.
85. Cravey, R.H. and Jain, N.C., Current status of blood alcohol methods. *J. Chromatogr. Sci.,* 12, 209, 1974.
86. Dubowski, K.M., Manual for Analysis of Alcohol in Biological Liquids. U.S. Department of Transportation Report DOT-TSC-NHTSA-76-4, 1977.
87. Machata, G., Über die gaschromatographische Blutalkoholbestimmung Analyse der Dampfphase. *Microchim. Acta,* 262, 1964.
88. Anthony, R.M., Suthejmer, C.A., and Sunshine, I., Acetaldehyde, methanol, and ethanol analysis by headspace gas chromatography. *J. Anal. Toxicol.,* 4, 43, 1980.
89. Christmore, D.S., Kelly, R.C., and Doshier, L.A., Improved recovery and stability of ethanol in automated headspace analysis. *J. Forensic. Sci.,* 29, 1038, 1984.
90. Watts, M.T. and McDonald, O.L., The effect of specimen type on the gas chromatographic headspace analysis of ethanol and other volatile compounds. *Am. J. Clin. Pathol.,* 87, 79, 1987.
91. Watts, M.T. and McDonald, O.L., The effect of sodium chloride concentration, water content, and protein on the gas chromatographic headspace analysis of ethanol in plasma. *Am. J. Clin. Pathol.,* 93, 357, 1990.
92. Smalldon, K.W. and Brown, G.A., The stability of ethanol in stored blood samples. II. The mechanism of ethanol oxidation. *Anal. Chim. Acta,* 66, 285, 1973.
93. Macchia, T., Mancinelli, R., Gentilli, S., Lugaresi, E.C., Raponi, A., and Taggi, F., Ethanol in biological fluids: headspace GC measurement. *J. Anal. Toxicol.,* 19, 241, 1995.
94. Zilly, M., Langmann, P., Lenker, U., Satzinger, V., et al., Highly sensitive gas chromatographic determination of ethanol in human urine samples. *J. Chromatogr. B,* 798, 179, 2003.
95. Iffland, R. and Jones, A.W., Evaluating alleged drinking after driving — the hip-flask defence. II. Congener analysis. *Med. Sci. Law,* 43, 39, 2003.
96. Krause, D. and Wehner, H.D., Blood alcohol/congeners of alcoholic beverages. *Forensic Sci. Int.,* 144, 177, 2004.
97. Bonte, W., *Begleistoffe alkoholischer Getränke,* Arbeitsmethoden der medizinischen und naturwissenschaftlichen Kriminalistik, Bd 17, Schmidt-Römhild, Lübeck, 1989.
98. Lachenmeier, D.W. and Musshoff, F., Begleitstoffgehalte alkoholischer Getränke Verlaufskontrollen, Chargenvergleich und aktuelle Konzentrationsbereiche. *Rechtsmedizin,* 14, 454, 2004.
99. Zuba, D., Parczewski, A., and Reichenbacher, M., Optimization of solid-phase microextraction conditions for gas chromatographic determination of ethanol and other volatile compounds in blood. *J. Chromatogr. B,* 773, 75, 2002.
100. De Martinis, B.S. and Martin, C.C., Automated headspace solid-phase micro-extraction and capillary gas chromatography analysis of ethanol in postmortem specimens. *Forensic Sci. Int.,* 128, 115, 2002.
101. Kühnholz, B. and Bonte, W., Methodische Untersuchungen zur Verbesserung des fuselalkoholnachweises in Blutproben. *Blutalkohol,* 20, 399, 1983.
102. Logan, B.K., Analysis of alcohol and other volatiles, in *Gas Chromatography in Forensic Science,* Tebbett, J., Ed., Elsevier, Amsterdam, 1992, chap. 4.
103. Tagliaro, F., Lubli, G., Ghielmi, S., Franchi, D., and Marigo, M., Chromatographic methods for blood alcohol determination. *J. Chromatogr.,* 580, 161, 1992.
104. Seto, E., Determination of volatile substances in biological samples by headspace gas chromatography. *J. Chromatogr.,* 674, 25, 1994.
105. Purdon, E.A., Distinguishing between ethanol and acetonitrile using gas chromatography and modified Widmark methods. *J. Anal. Toxicol.,* 17, 63, 1993.

106. Streete, P.J., Ruprah, M., Ramsey, J.D., and Flanagan, R.J., Detection and identification of volatile substances by headspace capillary gas chromatography to aid the diagnosis of acute poisoning. *Analyst,* 117, 1111, 1992.
107. Zuba, D., Piekoszewski, W., Pach, J., Winnik, L., and Parczewski, A., Concentrations of ethanol and other volatile compounds in the blood of acutely poisoned alcoholics. *Alcohol,* 26, 17, 2002.
108. Jones, A.W., A rapid method for blood alcohol determination by headspace analysis using an electrochemical detector. *J. Forensic Sci.,* 23, 283, 1978.
109. Dubowski, K.M., Method for alcohol determination in biological liquids by sensing with a solid-state detector. *Clin. Chem.,* 22, 863, 1976.
110. Criddle, W.J., Parry, K.W., and Jones, T.P., Determination of ethanol in alcoholic beverages using headspace procedure and fuel cell sensor. *Analyst,* 111, 507, 1986.
111. Kricka, L.J. and Thorpe, G.H.G., Immobilized enzymes in analysis. *Trends Biotechnol.,* 4, 253, 1986.
112. Varadi, M. and Adanyi, N., Application of biosensors with amperometric detection for determining ethanol. *Analyst,* 119, 1843, 1994.
113. Blaedel, W.J. and Engström, R.C., Reagentless enzyme electrodes for ethanol, lactate, and maleate. *Anal. Chem.,* 52, 1691, 1980.
114. Cheng, F.S. and Christian, G.D., Enzymatic determination of blood ethanol, with amperometric measurement of rate of oxygen depletion. *Clin. Chem.,* 24, 621, 1978.
115. Gulberg, E.L. and Christian, G.D., The use of immobilized alcohol oxidase in the continuous flow determination of ethanol with an oxygen electrode. *Anal. Chim. Acta,* 123, 125, 1981.
116. Gibson, T.D. and Woodward, J.R., Automated determination of ethanol using the enzyme alcohol oxidase. *Anal. Proc. Chem. Soc.,* 23, 360, 1986.
117. Gullbault, G.G., Danielsson, B., Mandenius, C.F., and Mosbach, K., Enzyme electrode and thermistor probes for determination of alcohols with alcohol oxidase. *Anal. Chem.,* 55, 1582, 1983.
118. Cenas, N., Rozgaite, J., and Kulys, J., Lactate, pyruvate, ethanol, and glucose-6-phosphate determination by enzyme electrode. *Biotech. Bioeng.,* 26, 551, 1984.
119. Gallignani, M., Garrigues, S., and Guardia, de la M., Derivative Fourier transform infrared spectrometric determination of ethanol in beer. *Analyst,* 119, 1773, 1994.
120. Ojanperä, I., Hyppölä, R., and Vuori, E., Identification of volatile organic compounds in blood by purge and trap PLOT-capillary gas chromatography coupled with Fourier transform infrared spectroscopy. *Forensic Sci. Int.,* 80, 201, 1996.
121. Pappas. A.A., Thompson, J.R., Porter, W.H., and Gadsden, R.H., High resolution proton nuclear magnetic resonance spectroscopy in the detection and quantitation of ethanol in human serum. *J. Anal. Toxicol.,* 17, 230, 1993.
122. Monaghan, M.S., Olsen, K.M., Ackerman, B.H., Fuller, G.L., Porter, W.H., and Pappas, A.A., Measurement of serum isopropanol and acetone metabolite by proton nuclear magnetic resonance: application to pharmacokinetic evaluation in a simulated overdose model. *Clin. Toxicol.,* 33, 141, 1995.
123. Robinson, A.G. and Loeb, J.N., Ethanol ingestion — commonest cause of elevated plasma osmolality. *N. Engl. J. Med.,* 284, 1253, 1971.
124. Hoffman, R.S., Smilkstein, M.J., Howland, M.A., and Goldfrank, L.R., Osmol gaps revisited: normal values and limitations. *Clin. Toxicol.,* 31, 81, 1993.
125. Osterloh, J.D., Kelly, T.J., Khayam-Bashi, H., and Romeo, R., Discrepancies in osmolal gaps and calculated alcohol concentrations. *Arch. Pathol. Lab. Med.,* 120, 637, 1996.
126. Purssell, R.A., Lynd, L.D., and Koga, Y., The use of osmole gap as a screening test for the presence of exogenous substances. *Toxicol. Rev.,* 23, 189, 2004.
127. Koga, Y., Purssell, R.A., and Lynd, L.D., The irrationality of the present use of the osmole gap: applicable physical chemistry principles and recommendations to improve the validity of current practices. *Toxicol. Rev.,* 23, 203, 2004.
128. Amato, I., Race quickens for non-stick blood monitoring technology. *Science,* 258, 892, 1992.
129. Ridder, T.D., Hendee, S.P., and Brown, C.D., Noninvasive alcohol testing using diffuse reflectance near-infrared spectroscopy. *Appl. Spectrosc.,* 59, 181, 2005.
130. Pan, S., Chung, H., Arnold, M.A., and Small, G.W., Near-infrared spectroscopic measurement of physiological glucose levels in variable matrices of protein and triglycerides. *Anal. Chem.,* 68, 1124, 1996.

131. Bonnichsen, R.K. and Ryhage, R., Determination of ethyl alcohol by computerized mass spectrometry. *Z. Rechsmed.*, 71, 134, 1972.
132. Jones, A.W., Mårdh, G., and Änggård, E., Determination of endogenous ethanol in blood and breath by gas chromatography-mass spectrometry. *Pharmacol. Biochem. Behav.*, 18(Suppl. 1), 267, 1983.
133. Dean, R.A., Thomasson, H.R., Dumaual, N., Amann, D., and Li, T.K., Simultaneous measurement of ethanol and ethyl-d5 alcohol by stable isotope gas chromatography-mass spectrometry. *Clin. Chem.*, 42, 367, 1996.
134. Takayasu, T., Ohshima, T., Tanaka, N., Maeda, H., Kondo, T., Nishigami, J., Ohtsuji, M., and Nagano, T., Experimental studies on postmortem diffusion of ethanol-d6 using rats. *Forensic Sci. Int.*, 76, 179, 1995.
135. Takayasu, T., Ohshima, T., Tanaka, N., Maeda, H., Kondo, T., Nishigami, J., and Nagano, T., Postmortem degradation of administered ethanol-d6 and production of endogenous ethanol experimental studies using rats and rabbits. *Forensic Sci. Int.*, 76, 129, 1995.
136. Ammon, E., Schäfer, C., Hofmann, U., and Klotz, U., Disposition and first-pass metabolism of ethanol in humans: is it gastric or hepatic and does it depend on gender. *Clin. Pharmacol. Ther.*, 59, 503, 1996.
137. Dubowski, K.M., The Technology of Breath-Alcohol Analysis, U.S. Department of Health and Human Services, DHHS Publ. (ADM) 92-1728, 1992.
138. Wilson, H.K., Breath-analysis; physiological basis and sampling techniques. *Scand. J. Work Environ. Health,* 12, 174, 1986.
139. Jones, A.W., Physiological aspects of breath-alcohol analysis. *Alc. Drug Driving,* 6, 1, 1990.
140. Mason, M. and Dubowski, K.M., Breath-alcohol analysis: uses, methods and some forensic problems — review and opinion. *J. Forensic Sci.,* 21, 9, 1976.
141. Harding, P., Breath-alcohol methods, in *Medicolegal Aspects of Alcohol,* Garriott, J., Ed., Lawyers & Judges, Tucson, 2004.
142. Gullberg, R.G., Breath alcohol analysis, in *Encyclopedia of Forensic and Legal Medicine,* Payne-James, J., Byard, R.W., Corey, T.S., and Henderson, C., Eds., Elsevier Science, Amsterdam, 2005, 21.
143. Manolis, A., The diagnostic potential of breath analysis. *Clin. Chem.,* 29, 5, 1983.
144. Phillips, M., Breath tests in medicine. *Sci. Am.,* 270, 52, 1992.
145. Jain, N.C. and Cravey. R.H., A review of breath alcohol methods. *J. Chromatogr. Sci.,* 12, 214, 1974.
146. Moynham, A., Perl, J., and Starmer, G.A., Breath-alcohol testing. *J. Traffic Med.,* 18, 167, 1990.
147. Jones, A.W., Pharmacokinetics of ethanol in saliva; comparison with blood and breath alcohol profiles, subjective feelings of intoxication and diminished performance. *Clin. Chem.,* 39, 1837, 1993.
148. Jones, A.W., Measurement of alcohol in blood and breath for legal purposes, in *Human Metabolism of Alcohol*, Crow, K.E. and Batt, R.D., Eds., CRC Press, Boca Raton, FL, 1989, 71.
149. Evans, R.P. and McDermott, F.T., Use of an Alcolmeter in a casualty department. *Med. J. Aust.,* 1, 1032, 1977.
150. Gibbs, K.A., Johnston, C.C., and Martin, S.D., Accuracy and usefulness of a breath-alcohol analyzer. *Ann. Emerg. Med.,* 13, 516, 1984.
151. Falkensson, M., Jones, A.W., and Sörbö, B., Bedside diagnosis of alcohol intoxication with a pocket-size breath-alcohol device sampling from unconscious subjects and specificity for ethanol. *Clin. Chem.,* 35, 918, 1989.
152. Frank, J.F. and Flores, A.L., The likelihood of acetone interference in breath alcohol measurement. *Alc. Drugs Driving*, 3, 1, 1987.
153. Jones, A.W. and Andersson, L., Biotransformation of acetone to isopropanol observed in a motorist involved in a sobriety check. *J. Forensic Sci.,* 40, 686, 1995.
154. Emerson, V.J., Holleyhead, R., Isaacs, D.J., Fuller, N.A., and Hunt, D.J., The measurement of breath alcohol. *J. Forensic Sci. Soc.,* 20, 1, 1980.
155. Harte, R.A., An instrument for the determination of ethanol in breath in law enforcement practice. *J. Forensic Sci.,* 16, 167, 1971.
156. Fransson, M., Jones, A.W., and Andersson, L., Laboratory evaluation of a new evidential breath-alcohol analyzer designed for mobile testing. *Med. Sci. Law,* 45, 61, 2005.
157. Caddy, G.R., Sobell, M.B., and Sobell, L.C., Alcohol breath tests: criterion times for avoiding contamination by mouth alcohol. *Behav. Res. Methods Instrum.,* 10, 814, 1978.
158. Gullberg, R.G., The elimination rate of mouth alcohol. Mathematical modeling and implications in breath alcohol analysis. *J. Forensic Sci.,* 37, 1363, 1992.

159. Langille, R.M. and Wigmore, J.G., The mouth alcohol effect after a mouthful of beer under social conditions. *Can. J. Forensic Sci. Soc.*, 33, 193, 2000.
160. Buczer, Y. and Wigmore, J.G., The significance of breath sampling frequency on the mouth alcohol effect. *Can. J. Forensic Sci. Soc.*, 35, 185, 2002.
161. Modell, J.G., Taylor, J.P., and Lee, J.Y., Breath alcohol values following mouthwash use. *J. Am. Med. Assoc.*, 270, 2955, 1993.
162. Jones, A.W., Reflections on the GERD defense. *DWI J. Law Sci.*, 20, 3, 2005.
163. Harding, P.M., Laessig, R.H., and Field, P.H., Field performance of the Intoxilyzer 5000: a comparison of blood- and breath-alcohol results in Wisconsin drivers. *J. Forensic Sci.*, 35, 1022, 1990.
164. Jones, A.W. and Andersson, L., Comparison of ethanol concentrations in venous blood and end-expired breath during a controlled drinking experiment. *Forensic Sci. Int.*, 132, 18, 2003.
165. Jones, A.W., Variability of the blood-breath alcohol ratio *in vivo*. *J. Stud. Alcohol,* 39, 1931, 1978.
166. Jones, A.W., Medicolegal alcohol determination — blood and/or breath-alcohol concentration. *Forensic Sci. Rev.,*12, 23, 2000.
167. Jones, A.W., Lindberg, L., and Olsson, S.G., Magnitude and time course of arterio-venous differences in blood-alcohol concentration in healthy men. *Clin. Pharmacokinet.,* 43, 1157, 2004.
168. Martin E., Moll, M., Schmid P., et al., The pharmacokinetics of alcohol in human breath, venous and arterial blood after oral ingestion. *Eur. J. Clin. Pharmacol.,* 26, 619, 1984.
169. Dubowski, K.M., Quality assurance in breath-alcohol analysis. *J. Anal. Toxicol.,* 18, 306, 1994.
170. Gullberg, R.G., Methodology and quality assurance in forensic breath alcohol analysis. *Forensic Sci. Rev.,* 12, 23, 2000.
171. Taylor, J.K., *Quality Assurance of Chemical Measurements*, Lewis, Chelsea, MI, 1987.
172. Jones, A.W., Edman-Falkensson, M., and Nilsson, L., Reliability of blood alcohol determinations at clinical chemistry laboratories in Sweden. *Scand. J. Clin. Lab. Invest.,* 35, 463, 1995.
173. Jones, A.W. and Fransson, M., Blood analysis by headspace gas chromatography: does a deficient sample volume distort ethanol concentrations? *Med. Sci. Law,* 43, 241, 2003.
174. Miller, B.A., Day, S.M., Vasquest, T.E., and Evans, F.M., Absence of salting out effects in forensic blood alcohol determination at various concentrations of sodium fluoride using semi-automated headspace gas chromatography. *Sci. Justice,* 44, 73, 2004.
175. Narayanan, S., The preanalytical phase — an important component of laboratory medicine. *Am. J. Clin. Pathol.,* 113, 429, 2000.
176. Matlow, A.G. and Berte, L.M., Sources of error in laboratory medicine. *Lab. Med.,* 35, 331, 2004.
177. Riley, D., Wigmore, J.G., and Yen, B., Dilution of blood collected for medicolegal alcohol analysis by intravenous fluids. *J. Anal. Toxicol.,* 20, 330, 1996.
178. Gullberg, R.G., The application of control charts in breath alcohol measurement systems. *Med. Sci. Law,* 33, 33, 1993.
179. Paulson, R. and Wachtel, M., Using quality control charts for quality assurance. *Lab. Med.,* 26, 409, 1995.
180. Winek, T., Winek, C.L., and Wahba, W.W., The effect of storage at various temperatures on blood alcohol concentration. *Forensic Sci. Int.,* 78, 179, 1996.
181. Meyer, T., Monge, P.K., and Sakshaug, J., Storage of blood samples containing alcohol. *Acta Pharmacol. Toxicol.,* 45, 282, 1979.
182. Dubowski, K.M., Gadsden, R.H., and Poklis, A., The stability of ethanol in human whole blood controls: an interlaboratory evaluation. *J. Anal. Toxicol.,* 21, 486, 1997.
183. Burnett, D., *Understanding Accreditation in Laboratory Medicine*, ACD Venture Publications, London, 1996.
184. Jones, G.R., Accreditation — toxicology, in *Encyclopedia of Forensic and Legal Medicine*, Payne-James, J., Byard, R.W., Corey, T.S., and Henderson, C., Eds., Elsevier Science, Amsterdam, 2005, 13.
185. Wilson, J.F. and Barnett, K., External quality assessment of techniques for assay of serum ethanol. *Ann. Clin. Biochem.,* 32, 540, 1993.
186. Jones, A.W., Biochemistry and physiology of alcohol: applications to forensic science and toxicology, in *Medicolegal Aspects of Alcohol,* Garriott, J., Ed., Lawyers & Judges, Tucson, 1996, 85.
187. Jones, A.W., Disposition and fate of ethanol in the body, in *Medicolegal Aspects of Alcohol,* 4th ed., Garriott, J., Ed., Lawyers & Judges, Tucson, 2003, 85.
188. Sedman, A.J., Wilkinson, P.K., Sakmar, E., Weidler, D.J., and Wagner, J.G., Food effect on absorption and metabolism of ethanol. *J. Stud. Alcohol,* 37, 1197, 1976.

189. Jones, A.W. and Jönsson, K.Å., Food-induced lowering of blood alcohol profiles and increased rate of elimination immediately after a meal. *J. Forensic Sci.*, 39, 1084, 1994.
190. Jones, A.W., Inter-individual variations in the disposition and metabolism of ethanol in healthy men. *Alcohol,* 1, 385, 1984.
191. Jones, A.W., Quantitative relationships among ethanol concentrations in blood, breath, saliva, and urine during ethanol metabolism in man, in *Proc. 8th Int. Conf. Alcohol, Drugs, and Traffic Safety,* Goldberg, L., Ed., Almqvist & Wiksell, Stockholm, 1981, 550.
192. Schmitt, G., Aderjan, R., Keller, T., and Wu, M., Ethyl glucuronide: an unusual ethanol metabolite in humans, synthesis, and analytical data. *J. Anal. Toxicol.,* 19, 91, 1995.
193. Sato, N. and Kitamura, T., First-pass metabolism of ethanol: a review. *Gastroenterology,* 111, 1143, 1996.
194. Park, B.K., Pirmohamed, M., and Kitteringham, N.R., The role of cytochrome P450 enzymes in hepatic and extrahepatic human drug toxicity. *Pharmacol. Ther.,* 68, 385, 1995.
195. Teschke, P. and Gellert, J., Hepatic microsomal ethanol oxidizing systems (MEOS): metabolic aspects and clinical implications. *Alcohol Clin. Exp. Res.,* 10, 20S, 1986.
196. Lieber, C.S., The discovery of the microsomal ethanol oxidizing system and its physiologic and pathologic role. *Drug Metab. Rev.,* 36, 511, 2004.
197. Jones, A.W., Disappearance rate of ethanol from the blood of human subjects; implications in forensic toxicology. *J. Forensic Sci.,* 38, 104, 1993.
198. Lieber, C.S., Mechanisms of ethanol induced hepatic injury. *Pharmacol. Ther.,* 46, 1, 1990.
199. Lee, W.M., Drug-induced hepatotoxicity. *N. Engl. J. Med.,* 333, 1118, 1995.
200. Jones, A.W., Alcohol and drug interactions, in *Handbook of Drug Interactions: A Clinical and Forensic Guide,* Mozayani, A. and Raymond, L.P., Eds., Humana Press, Totowa, NJ, 2003, 395.
201. Hu, Y., Ingelman-Sundberg, M., and Lindros, K.O., Inductive mechanisms of cytochrome P450 2E1 in liver: interplay between ethanol treatment and starvation. *Biochem. Pharmacol.,* 50, 155, 1995.
202. Slattery, J.T., Nelson, S.D., and Thummel, K.E., The complex interaction between ethanol and acetaminophen. *Clin. Pharm. Ther.,* 60, 241, 1996.
203. Ahmed, F.E., Toxicological effects of ethanol on human health. *C. R. Toxicol.,* 25, 347, 1995.
204. Oneta, C.M., Lieber, C.S., Li, J., Ruttimann, S., et al., Dynamics of cytochrome P4502E1 activity in man: induction by ethanol and disappearance during withdrawal. *J. Hepatol.,* 36, 47, 2002.
205. Nevo, I. and Hamon, M., Neurotransmitter and neuromodulatory mechanisms involved in alcohol abuse and alcoholism. *Neurochem. Int.,* 26, 305, 1995.
206. Mehta, A.K. and Ticka, M.K., An update on $GABA_A$ receptors. *Brain Res. Rev.,* 29, 196, 1999.
207. Bormann, J., The ABC of GABA receptors. *TIPS,* 21, 16, 2000.
208. Krystal, J.H., Petrakis, I.L., Mason, G., Trevisan, L., and D'Souza, D.C., *N*-Methyl-D-aspartate glutamate receptors and alcoholism: reward, dependence, treatment, and vulnerability. *Pharm. Ther.,* 99, 79, 2003.
209. Widmark, E.M.P., *Die theoretischen Grundlagen und die praktische Verwendbarkeit der gerichtlich-medizinischen Alkoholbestimmung*, Urban und Schwarzenberg, Berlin, 1932, 1–140.
210. Kalant, H., Current state of knowledge about the mechanisms of alcohol tolerance. *Addiction Biol.,* 1, 133, 1996.
211. Hoffman, P.L. and Tabakoff, B., Alcohol dependence: a commentary on mechanisms. *Alc. Alcohol,* 31, 333, 1996.
212. Bogen, E., The human toxicology of alcohol, in *Alcohol and Man,* Emerson, H., Ed., Macmillan, New York, 1932, chap. 6.
213. Penttilä, A. and Tenhu, M., Clinical examination as medicolegal proof of alcohol intoxication. *Med. Sci. Law,* 16, 95, 1976.
214. Kataja, M., Penttilä, A., and Tenhu, M., Combining the blood alcohol and clinical examination for estimating the influence of alcohol. *Blutalkohol,* 12, 108, 1975.
215. Jones, A.W. and Holmgren, P., Comparison of blood-ethanol concentrations in deaths attributed to acute alcohol poisoning and chronic alcoholism. *J. Forensic Sci.,* 48, 874, 2003.
216. Koski, A., Ojanpera, I., and Vuori, E., Alcohol and benzodiazepines in fatal poisonings. *Alcohol Clin. Exp. Res.,* 26, 956, 2002.
217. Jones, B.M. and Vega, A., Fast and slow drinkers; blood alcohol variables and cognitive performance. *J. Stud. Alc.,* 34, 797, 1973.

218. Moskowitz, H. and Burns, M., Effects of rate of drinking on human performance. *J. Stud. Alcohol*, 37, 598, 1976.
219. Jones, A.W. and Neri, A., Age-related differences in blood-ethanol parameters and subjective feelings of intoxication in healthy men. *Alc. Alcohol,* 20, 45, 1985.
220. Zink, P. and Reinhardt, G., Der Verlauf der Blutalkoholkurve bei großen Trinkmengen. *Blutalkohol,* 21, 422, 1984.
221. Wilkinson, P.K., Pharmacokinetics of ethanol. *Alcohol Clin. Exp. Res.,* 4, 6, 1980.
222. Holford, N.H.G., Clinical pharmacokinetics of ethanol. *Clin. Pharmacokinet.,* 13, 273, 1987.
223. Ludden, T.M., Nonlinear pharmacokinetics — clinical implications. *Clin. Pharmamcokinet.,* 20, 429, 1991.
224. Roland, M. and Tozer, T.N., *Clinical Pharmacokinetics: Concepts and Applications.* Lea & Febiger, Philadelphia, 1980.
225. Jones, A.W., Forensic science aspects of ethanol metabolism, in *Forensic Science Progress*, Mahley, A. and Williams, R.L., Eds., Springer Verlag, New York, 1991, 33.
226. Andreasson, R. and Jones, A.W., The life and work of Erik M.P. Widmark. *Am. J. Forensic Med. Pathol.,* 17, 177, 1996.
227. Jones, A.W. and Andersson, L., Influence of age, gender, and blood-alcohol concentration on rate of alcohol elimination from blood in drinking drivers. *J. Forensic Sci.,* 41, 922, 1996.
228. Jones, A.W. and Sternebring, B., Kinetics of ethanol and methanol in alcoholics during detoxification. *Alc. Alcohol,* 27, 647, 1992.
229. Bogusz, M., Pach, J., and Stasko, W., Comparative studies on the rate of ethanol elimination in acute poisoning and in controlled conditions. *J. Forensic Sci.,* 22, 446, 1977.
230. Keiding, S., Christensen, N.J., Damgaard, S.E., Dejgård, A., et al., Ethanol metabolism in heavy drinkers after massive and moderate alcohol intake. *Biochem. Pharmacol.,* 20, 3097, 1983.
231. Haffner, H.T., Besserer, K., Stetter, F., and Mann, K., Die Äthanol-Eliminations-geschwindigkeit bei Alkoholikern unter besonderer Berücksichtigung der Maximalwertvarianyte der forensischsen BAK-Rückrechnung. *Blutalkohol,* 28, 46, 1991.
232. Conney, A.H., Induction of drug-metabolizing enzymes: a path to the discovery of multiple cytochromes P450. *Annu. Rev. Pharmacol.,* 43, 1, 2003.
233. Yap, M., Mascord, D.J., Starmer, G.A., and Whitfield, J.B., Studies on the chrono-pharmacokinetics of ethanol. *Alc. Alcohol,* 28, 17, 1993.
234. Lötterle, J., Husslein, E.M., Bolt, J., and Wirtz, P.M., Tageszeitliche Unterschiede der Alkoholresorption. *Blutalkohol,* 26, 369, 1989.
235. Johnson, R.D., Horowitz, M., Maddox, A.F., Wishart, J.M., and Shearman, D.J.C., Cigarette smoking and rate of gastric emptying: effect on alcohol absorption. *Br. Med. J.,* 302, 20, 1991.
236. Watson, P.E., Watson, I.D., and Batt, R.D., Prediction of blood alcohol concentration in human subjects; updating the Widmark equation. *J. Stud. Alcohol*, 42, 547, 1981.
237. Gullberg, R.G. and Jones, A.W., Guidelines for estimating the amount of alcohol consumed from a single measurement of blood alcohol concentration; re-evaluation of Widmark's equation. *Forensic Sci. Int.,* 69, 119, 1994.
238. Oneta, C.M., Pedrosa, M., Ruttimann, S., Russell, R.M., and Seitz, H.K., Age and bioavailability of alcohol. *Z. Gastroenterol.,* 39, 783, 2001.
239. Tozer, T.N. and Rubin, G.M., Saturable kinetics and bioavailability determination, in *Pharmacokinetics; Regulatory, Industrial, Academic Perspectives.* Vol. 33, *Drugs and the Pharmaceutical Sciences*, Welling, P.G. and Tse, F.L.S., Eds., Marcel Dekker, New York, 1988, 473.
240. Lundquist, F. and Wolthers, H., The kinetics of alcohol elimination in man. *Acta Pharmacol. Toxicol.,* 14, 265, 1958.
241. Wilkinson, P.K., Sedman, A.J., Sakmar, E., Kay, D.R., and Wagner, J.G., Pharmacokinetics of ethanol after oral administration in the fasting state. *J. Pharmacokinet. Biopharm.,* 5, 207, 1977.
242. Wagner, J.G., Wilkinson, P.K., and Ganes, D.A., Parameters V_m and k_m for elimination of alcohol in young male subjects following low doses of alcohol. *Alc. Alcohol,* 24, 555, 1989.
243. Lewis, M.J., Blood alcohol: the concentration-time curve and retrospective estimation of level. *J. Forensic Sci. Soc.,* 26, 95, 1985.
244. Forrest, A.R.W., Non-linear kinetics of ethyl alcohol metabolism. *J. Forensic Sci. Soc.,* 26, 121, 1986.
245. Bosron, W.F. and Li, T.K., Genetic polymorphism of human liver alcohol and aldehyde dehydrogenases, and their relationship to alcohol metabolism and alcoholism. *Hepatology,* 6, 502, 1986.

246. Li, T.K., Yin, S.J., Crabb, D.W., O'Connor, S., and Ramchandani, V.A., Genetic and environmental influences on alcohol metabolism in humans. *Alcohol Clin. Exp. Res.,* 25, 136, 2001.
247. Oneta, C.M., Simanowski, U.A., Martinez, M., Allali-Hassani, A., et al., First pass metabolism of ethanol is strikingly influenced by the speed of gastric emptying. *Gut,* 43, 612, 1998.
248. Gentry, R.T., Baraona, E., and Lieber, C.S., Gastric first pass metabolism of alcohol. *J. Lab. Clin. Med.,* 123, 21, 1994.
249. Fressa, M., DiPadova, C., Pozzato, G., Terpin, M., Baraona, E., and Lieber, C.S., High blood alcohol levels in women; the role of decreased gastric alcohol dehydrogenase activity and first pass metabolism. *N. Engl. J. Med.,* 322, 95, 1990.
250. Levitt, M.D., The case against first-pass metabolism of ethanol in the stomach. *J. Lab. Clin. Med.,* 123, 28, 1994.
251. Levitt, M.D., Lack of clinical significance of the interaction between H_2-receptor antagonists and ethanol. *Aliment. Pharmacol. Ther.,* 7, 131, 1993.
252. Levitt, M.D. and Levitt, D.G., The critical role of the rate of ethanol absorption in the interpretation of studies purporting to demonstrate gastric metabolism of ethanol. *J. Pharmacol. Ther.,* 269, 297, 1994.
253. Arora, S., Baraona, E., and Lieber, C.S., Alcohol levels are increased in social drinkers receiving rantitidne. *Am. J. Gastroenterol.,* 95, 208, 2000.
254. DiPadova, C., Roine, R., Fressa, M., Gentry, T.R., Baraona, E., and Lieber, C.S., Effects of ranitidine on blood alcohol levels after ethanol ingestion. *J. Am. Med. Assoc.,* 267, 83, 1992.
255. Roine, R., Gentry, T., Hernandez-Munoz, R., et al., Aspirin increases blood alcohol concentration in humans after ingestion of alcohol. *J. Am. Med. Assoc.,* 264, 2406, 1990.
256. Amir, I., Anwar, N., Baraona, E., and Lieber, C.S., Ranitidine increases the bioavailability of imbibed alcohol by accelerating gastric emptying. *Life Sci.,* 58, 511, 1996.
257. Palmer, R.H., Frank, W.O., Nambi, P., Wetherington, J.D., and Fox, M.J., Effects of various concomitant medications on gastric alcohol dehydrogenase and its first-pass metabolism of ethanol. *Am. J. Gastroenterol.,* 86, 1749, 1991.
258. Lieber, C.S., Alcohol and the liver: 1994 update. *Gastroenterology,* 106, 1085, 1994.
259. Seitz., H.K., Egerer, G., Simanowski, U.A., Eckey, R., Agarwal, D.P., Goedde, H.W., and Von Wartburg, J.P., Human gastric alcohol dehydrogenase activity: effect of age, sex, and alcoholism. *Gut,* 34, 1433, 1993.
260. Raufman, J.P., Notar-Francesco, V., Raffaniello, R.D., and Straus, E.W., Histamine-2 receptor antagonists do not alter serum ethanol levels in fed, nonalcoholic men. *Ann. Intern. Med.,* 118, 488, 1983.
261. Kendall, M.J., Spannuth, F., Wait, R.P., Gibson, G.J., et al., Lack of effect of H2-receptor antagonists on the pharmacokinetics of alcohol consumed after food at lunchtime. *Br. J. Clin. Pharmacol.,* 37, 371, 1994.
262. Toon, S., Khan, A.Z., Holt, B.I., Mullins, F.G.P., Langley, S.J., and Rowland, M.M., Absence of effect of ranitidine on blood alcohol concentrations when taken morning, midday, or evening with or without food. *Clin. Pharmacol. Ther.,* 55, 385, 1994.
263. Bye, A., Lacey, L.F., Gupta, S., and Powell, J.R., Effect of ranitidine hydrochloride (150 mg twice daily) on the pharmacokinetics of increasing doses of ethanol (0.15, 0.3, 0.6 g/kg). *Br. J. Clin. Pharmacol.,* 41, 129, 1996.
264. Jönsson, K.Å., Jones, A.W., Boström, H., and Andersson, T., Lack of effect of omeprazole, cimetidine, and ranitidine, on the pharmacokinetics of ethanol in fasting male volunteers. *Eur. J. Clin. Pharmacol.,* 42, 209, 1992.
265. Melander, O., Liden, A., and Melander, A., Pharmacokinetic interaction of alcohol and acetylsalicylic acid. *Eur. J. Clin. Pharmacol.,* 48, 151, 1995.
266. Passananti, G.T., Wolff, C.A., and Vesell, E.S., Reproducibility of individual rates of ethanol metabolism in fasting subjects. *Clin. Pharmacol. Ther.,* 47, 389, 1990.
267. Fraser, A.G., Rosalki, S.B., Gamble, G.D., and Pounder, R.E., Inter-individual and intra-individual variability of ethanol concentration-time profiles: comparison of ethanol ingestion before or after an evening meal. *Br. J. Clin. Pharmacol.,* 40, 387, 1995.
268. Jones, A.W. and Jönsson, K.Å., Between subject and within subject variations in the pharmacokinetics of ethanol. *Br. J. Clin. Pharmacol.,* 37, 427, 1994.
269. Welling, P.G., Lyons, L.L., Elliot, R., and Amidon, G.L., Pharmacokinetics of alcohol following single low doses to fasted and nonfasted subjects. *J. Clin. Pharmacol.,* 199, 1977.

270. Welling, P.G., How food and fluid affect drug absorption. *Postgrad. Med.,* 62, 73, 1977.
271. McFarlane, A., Pooley, L., Welch I., Rumsey, R.D.E., and Read, N.W., How does dietary lipid lower blood alcohol concentrations? *Gut,* 27, 15, 1986.
272. Winstanley, P.A. and Orme, M.L.E., The effects of food on drug bioavailability. *Br. J. Clin. Pharmacol.,* 28, 621, 1989.
273. Millar, K., Hammersley, R.H., and Finnigan, F., Reduction of alcohol-induced performance by prior ingestion of food. *Br. J. Psychol.,* 83, 261, 1992.
274. Jones, A.W., Jönsson K.Å., and Kechagias, S., Effect of high-fat, high-protein and high-carbohydrate meals on the pharmacokinetics of a small dose of alcohol. *Br. J. Clin. Pharmacol.,* 44, 521, 1997.
275. Ramachandani, V.A., Kwo, P.Y., and Li, T.K., Effect of food and food composition on alcohol elimination rates in healthy men and women. *J. Clin. Pharmacol.,* 41, 1345, 2001.
276. Ramachandani, V.A., Bosron, W.T., and Li, T.K., Research advances in ethanol metabolism. *Pathol. Biol.,* 49, 676, 2001.
277. Kalant, H., Effects of food and body composition on blood alcohol levels, in *Comprehensive Handbook of Alcohol Related Pathology*, Preedy, V.R. and Watson, R.R., Eds., Elsevier/Academic Press, Amsterdam, 2005, 87.
278. Hahn, R.G., Norberg, Å., Gabrielsson, J., Danielsson, A., and Jones, A.W., Eating a meal increases the clearance of ethanol given by intravenous infusion. *Alc. Alcohol,* 29, 673, 1994.
279. Pedrosa, M.C., Russell, R.M., Saltzman, J.R., Golner, B.B., Dallal, G.E., Sepe, T.E., Oats, E., Egerer, G., and Seitz, H.K., Gastric emptying and first-pass metabolism of ethanol in elderly subjects with and without atrophic gastritis. *Scand. J. Gastroenterol.,* 31,671, 1996.
280. Preedy, V.R. and Watson, R.R., Eds., *Comprehensive Handbook of Alcohol Related Pathology,* Elsevier/Academic Press, London, 2004.
281. O'Neal, C.L., Wolf, C.E., Levine, B., Kunsman, G., and Poklis, A., Gas chromatographic procedures for determination of ethanol in postmortem blood using *t*-butanol and methyl ethyl ketone as internal standard. *Forensic Sci. Int.,* 83, 31, 1996.
282. Hunsaker, D.M. and Hundsaker, J.C., III, Blood and body fluid analysis, in *Encyclopedia of Forensic and Legal Medicine,* Payne-James, J., Byard, R.W., Corey, T.S., and Henderson, C., Eds., Elsevier Science, Amsterdam, 2005, 29.
283. Wagner, J.G., Wilkinson, P.K., and Ganes, D.A., Estimation of the amount of alcohol ingested from a single blood alcohol concentration. *Alc. Alcohol,* 25, 379, 1990.
284. Kwo, P.Y., Ramchandani, V.A., O'Connor, S., Amann, D., Carr, L.G., Sandrasgaran, K., Kopecky, K.K., and Li, T.K., Gender differences in alcohol metabolism: relationship to liver volume and effect of adjusting for body composition. *Gastroenterology,* 115, 1552, 1998.
285. Li, T.K, Beard, J.D., Orr, W.E., Kwo, P.Y., Ramchandani, V.A., and Thomasson, H.R., Variation in ethanol pharmacokinetics and perceived gender and ethnic differences in alcohol elimination. *Alcohol Clin. Exp. Res.,* 24, 415, 2000.
286. Jackson, P.R., Tucker, G.T., and Woods, H.F., Backtracking booze with Bayes — the retrospective interpretation of blood alcohol data. *Br. J. Clin. Pharmacol.,* 31, 55, 1991.
287. Arora, S., Bararona, E., and Lieber, C.S., Alcohol levels are increased in social drinkers receiving ranitidine. *Am. J. Gastroenterol.,* 95, 208, 2000.
288. Norberg, Å., Jones, A.W., Hahn, R.G., and Gabrielsson, J., Role of variability in explaining ethanol pharmacokinetics. *Clin. Pharmacokinet.,* 42, 1, 2003.
289. Smith, G.D., Shaw, L.J., Maini, P.K., Ward, R.J., Peters, T.J., and Murray, J.D., Mathematical modelling of ethanol metabolism in normal subjects and chronic alcohol misusers. *Alc. Alcohol,* 28, 25, 1993.
290. Komura, S., Fujimiya, T., and Yoshimoto, K., Fundamental studies on alcohol dependence and disposition. *Forensic Sci. Int.,* 80, 99, 1996.
291. Fujimiya, T., Uemura, K., Ohbora, Y., and Komura, S., Problems in pharmacokinetic analysis of alcohol disposition: a trial of the Bayesian Least-Squares method. *Alcohol Clin. Exp. Res.,* 20, 2A, 1996.
292. Pieters, J. E., Wedel, M., and Schaafsma, G., Parameter estimation in a three-compartment model for blood alcohol curves. *Alc. Alcohol,* 25, 17, 1990.
293. Kosten, T.R. and O'Connor, P.G., Management of drug and alcohol withdrawal. *N. Engl. J. Med.,* 348, 1786, 2003.

CHAPTER 3

Post-Mortem Alcohol — Aspects of Interpretation

Derrick J. Pounder, M.D.[1] **and Alan Wayne Jones, D.Sc.**[2]
[1] Department of Forensic Medicine, University of Dundee, Scotland, U.K.
[2] Department of Forensic Toxicology, University Hospital, Linköping, Sweden

CONTENTS

3.1 INTRODUCTION

In post-mortem toxicology, the sample most commonly submitted for analysis is blood and the most frequently encountered substance is alcohol. The technical details and procedures for measuring ethanol in blood and other body fluids obtained from a corpse are essentially the same as those used when analyzing specimens from the living. However, the interpretation of the analytical results obtained from autopsy samples is confounded by problems such as the lack of homogeneity of blood samples, microbial alcohol production post-mortem, alcohol diffusion from gastric residue and contaminated airways, and the lack of or unreliability of information on the clinical condition of the person immediately prior to death. On the other hand, autopsy offers opportunities for sampling body fluids and tissues not accessible or not readily available in the living. Sampling of blood from multiple vascular sites, the vitreous humor of the eye, gastric contents, sequestered hematomas, as well as bile, brain, skeletal muscle, cerebrospinal fluid, and liver are all possible. Nevertheless, multiple sampling at autopsy can only partly compensate for the increased interpretative difficulties created by the various post-mortem confounding factors. As a result it is necessary to apply a greater degree of caution in the interpretation of post-mortem ethanol analyses and to take into account the totality of the available information, which should always include not only the results of the autopsy examination but also the scene of death examination and anamnestic data. A single autopsy blood ethanol concentration is commonly uninterpretable without concurrent vitreous humor and urine ethanol concentrations, as well as information gleaned from the scene of death and case history.

3.2 POST-MORTEM SPECIMENS FOR ALCOHOL ANALYSIS

Within a few hours of death the blood within the vascular system clots and simultaneously there is clot lysis. The effectiveness of the clot lysis will determine whether a blood sample obtained at autopsy is clotted, or completely fluid, or partly clotted and partly fluid. The fibrin clots invariably entrap large numbers of red blood cells so that the resulting clot is relatively red cell rich and serum poor. Occasionally the heart and great vessels may contain a large two-layered clot, the lower part typically red clot and the upper part pale yellow rubbery clot largely devoid of red cells (so called "chicken fat"). Consequently "blood samples" obtained at autopsy are variable in their red cell and protein content and this will have some influence on the measured ethanol concentration since ethanol is distributed in the water portion of blood. Blood obtained from limb vessels is most likely to be fluid and largely devoid of clots and therefore provides as homogeneous a sample for analysis as can be hoped for. The presence of blood clots will not necessarily have a negative influence on the accuracy of the blood alcohol analysis using headspace gas chromatography.[1]

Serum and plasma contains approximately 10 to 15% more water than whole blood. Since ethanol is distributed in the water portion of blood it can be expected that the concentration in plasma is approximately 10 to 15% higher than the corresponding whole blood concentration. This should be borne in mind whenever alcohol has been measured in serum or plasma in hospital prior to death or where a pre-mortem hospital sample of serum or plasma is subsequently analyzed for alcohol.

In 134 blood samples from healthy men and women[2] the mean blood alcohol concentration (BAC) was 105 mg/dL (range 22 to 155) and the mean serum alcohol concentration was 121 mg/dL (range 25 to 183) with a mean serum:blood ratio of 1.15 (range 1.10 to 1.25, standard deviation [SD] 0.02); 3 of the 134 serum:blood ratios were between 1.21 and 1.25. A larger study on 235 subjects[3] produced a similar mean serum:blood ratio of 1.14 with a range of 1.04 to 1.26 and a normal distribution with SD of 0.041. Ethanol concentration in red blood cells was reported in 167 of these subjects and ratio red cells:blood ethanol ranged from 0.66 to 1.00 with a mean of 0.865 and a negatively skewed distribution with an SD of 0.065. Given these data it is evident that there may be significant differences in ethanol concentrations between different blood samples obtained at the same time from the same corpse, since an autopsy "blood" sample may range in composition from being largely red cells at one extreme to largely plasma at the other. However, in practice, most autopsy blood samples will tend to be plasma rich rather than red blood cell rich because autopsy sampling procedures tend to avoid clots and favor clot-free fluid.

The water content of whole blood decreases post-mortem, and because ethanol is distributed only in the water phase of the body, this will cause the BAC to decrease. In a study of 71 cadavers[4] with a blood sample taken within 10 h of death (mean 2.1 h, range 0 to 9.6 h) the water content ranged between 72.4 and 89.3%, mean 80.4%, which is closely similar to the water content of blood from living persons (79.9 to 82.3% for women and 77.5 to 80.6% for men). Second samples taken from the same cadavers from 8 to 229 h post-mortem had a lower water content ranging between 64.4 and 88.0%, mean 74.0%. However, differences in blood ethanol concentration between the two sampling times were more strongly influenced by other post-mortem factors, such as putrefaction, rather than by water content changes, so that correcting a post-mortem blood alcohol for water content is not generally recommended.

3.3 BLOOD ALCOHOL CONCENTRATION

Ethanol is a central nervous system depressant drug exerting its effects in part via the $GABA_A$ inhibitory receptor. At very high BACs, stupor is followed by coma and then paralysis of the respiratory center in the brain stem so that breathing stops and death ensues. The lethal range of blood ethanol concentration is based on published case reports of human fatalities and the experimentally derived LD_{50} values of 500 to 550 mg/dL in rats, guinea pigs, chickens, and dogs.[5] However, a review of actual cases suggests that a concentration of 250 mg/dL might be potentially lethal rather than the higher figures commonly quoted.[6] In a retrospective review of 693 deaths attributed to acute alcohol poisoning, the mean BAC in femoral venous blood at autopsy was 360 mg/dL and the 5th and 95th percentiles were 220 and 500 mg/dL.[7] (For fatal blood ethanol levels in various series, see Niyogi.[8]) The often quoted lethal blood ethanol range of above 400 or 450 mg/dL may only apply to uncomplicated deaths as a result of acute alcohol poisoning in inexperienced drinkers. The complexity of interpreting the significance of high blood ethanol concentrations at autopsy is illustrated by a review[9] of fatalities with blood ethanol above 300 mg/dL. This study disclosed 502 cases attributable to acute ethanol poisoning alone but 24 resulting from well-documented natural causes, 260 from obvious trauma or violence, and 28 with a combination of a high blood ethanol and additional contributing or related abnormalities. An unusual form of sexual gratification involved self-administration of an alcohol-containing enema (klezmomania), which proved fatal with a BAC of 400 mg/dL found at autopsy following a wine enema.[10]

In a series of 115 deaths attributed to acute alcoholism 59% showed some asphyxial element, either postural asphyxia (positional asphyxia) or inhalation of vomit.[6] Since acute alcohol intoxication may induce vomiting and also suppresses the gag reflex, as part of its cerebral depressant effect, there is a high risk of inhalation of vomit. Alcohol-related deaths associated with an asphyxial element may show considerably lower BACs than uncomplicated alcohol poisonings. For this reason it is particularly important to have accurate documentation of the position of the body as found and any evidence of inhalation of vomitus at the scene of death, since passive regurgitation of gastric contents and contamination of the airways may occur post-mortem during removal of the body to the mortuary. In many of these deaths in which asphyxia is a contributing factor the urine alcohol is considerably higher than the blood alcohol suggesting that the mechanism of death was coma resulting from a high BAC and subsequent respiratory embarrassment and anoxia. In such fatalities the BAC observed at autopsy is not necessarily the same that caused death owing to ongoing metabolism up to the time of death.[6] The diagnostic features of accidental postural asphyxia are a body position that compromises breathing, such as abnormal neck flexion or limitation of chest movement, together with evidence of the accidental adoption of that position and an explanation for failure to escape, such as alcohol intoxication, in the absence of another explanation for the death.[11] Acute alcohol intoxication is a significant risk factor for this mode of death accounting for 22 of 30 cases in one study.[12]

A high autopsy blood ethanol concentration, although indicating chemical intoxication at the time of death, does not necessarily imply that there were observable clinical manifestations of drunkenness and this is particularly so in chronic alcoholics.[13] Alcohol abusers develop tolerance to alcohol to the extent that they can maintain very high blood ethanol concentrations in the potentially lethal range. In an Australian study[14] blood alcohol levels were determined in chronic alcoholics presenting to a detoxification service. Of the 32 subjects, all appeared affected by alcohol with 23 showing altered mood or behavior, 6 appearing confused, and 3 drowsy, but none was stuporous or comatose. All displayed ataxia and dysarthria of varying degrees. The blood ethanol concentration ranged from 180 to 450 mg/dL with a mean of 313 mg/dL and 26 of the 32 were above 250 mg/dL. A similar Swedish study[15] identified 24 patients who attended a hospital casualty department and were found to have blood ethanol concentrations above 500 mg/dL. Of the 16 patients for whom there were complete data available, 8 were either awake or could be aroused by nonpainful stimuli. All left the hospital alive within 24 h. It is suggested that this tolerance to high blood ethanol levels seen in chronic alcoholics is primarily the result of neuronal adaptation. Physical dependence on ethanol, as demonstrated by the development of withdrawal signs and symptoms on stopping drinking, similarly indicates the existence of an adaptation process.

There are anecdotal descriptions of alcoholics surviving remarkably high levels of blood alcohol. In one instance,[16] a 24-year-old female chronic alcoholic presented at the hospital with abdominal pain. She was agitated and slightly confused but alert, responsive to questioning, and orientated to person and place, though unclear as to time. Her serum ethanol was 1510 mg/dL, which corresponds to a concentration in whole blood of about 1313 mg/dL (1510/1.15). After 12 h treatment with intravenous fluids, electrolyte replacement, chlordiazepoxide, and intensive care monitoring, she felt well, and was symptomless at discharge 2 days later. Similarly, a 52-year-old, 66-kg male was found unconscious in a bar with a blood ethanol concentration of 650 mg/dL and survived with minimum treatment comprising protection against aspiration and the occasional use of oxygen.[17] A 23-year-old, 57-kg female chronic alcoholic, who was admitted to the hospital in a comatose state with a blood ethanol level of 780 mg/dL, had apparently consumed 390 mL of absolute alcohol in the form of a bottle of bourbon. She was discharged 11 h later with a blood ethanol level of 190 mg/dL; the disappearance of ethanol from her blood seemingly followed a logarithmic function.[18] Two further case reports describe more stormy clinical courses but with survival, one with a serum ethanol level of 1127 mg/dL,[19] another with a blood ethanol level of 1500 mg/dL.[20] One individual arrested in Sweden for drunk-driving had a BAC of 545 mg/dL.[21] This tolerance to high blood ethanol concentrations seen in chronic alcoholics makes it difficult to interpret the significance of a blood ethanol level obtained at autopsy from such a person.

On the other hand, nonlethal levels of ethanol may be of particular significance in some types of death. Ethanol adversely affects thermal regulation and, depending upon the ambient temperature, may cause either hypothermia or hyperthermia.[22] There is a large body of experimental and clinical data available regarding the hypothermic effect of alcohol on both animals and humans at different degrees of cold exposure. In deaths related to acute alcohol poisoning, a mean BAC of 170 mg/dL in hypothermia-associated deaths contrasted with a much higher mean BAC of 360 mg/dL in deaths without associated hypothermia.[23]Alcohol consumption accelerates body heat loss by inducing dilatation of peripheral blood vessels and relaxing muscles, thereby inhibiting shivering thermogenesis. Heat loss is further facilitated by behavioral effects consequent on a feeling of warmth and comfort, and central nervous system depression. At a biochemical level, alcohol may inhibit the protective acute ketogenesis induced by the effects of hypothermia and so shorten the survival time.[24] Deaths from hypothermia associated with sobriety or a BAC of less than 100 mg/dL have a higher frequency of the macroscopic autopsy signs of hypothermia — bright red lividity, violet patches over the knees and elbows, acute gastric mucosa erosions, pancreatitis, and paradoxical undressing — than cases with a BAC greater than 100 mg/dL, suggesting that intoxication is associated with more rapid death before visible pathological signs develop.[24] The importance of ethanol in hyperthermic deaths is less well appreciated but illustrated by Finnish sauna fatalities. In a series of 228 hyperthermic deaths (221 sauna related) alcohol had been consumed in 192 cases and the consumption was categorized as "heavy" in 61.[25] Similarly, complex cerebral dysfunctions induced by alcohol are thought to be significant in the syndrome of sudden, alcohol-associated, cranio-facial traumatic death.[26] In this syndrome individuals who have collapsed and died at the scene of an assault are found at autopsy to have facial trauma insufficient to account for the death together with a high but nonlethal BAC.

The simultaneous presence of another drug with ethanol further complicates the interpretation of the concentrations measured at autopsy. Investigations into drug–alcohol interactions are complex because these interactions are influenced by the timing of administration of alcohol relative to the drug and the specific dosages. As well as being oxidized by alcohol dehydrogenase (ADH), ethanol is metabolized to acetaldehyde by a microsomal enzyme, cytochrome P4502E1 (CYP2E1). This same enzyme is involved in oxidative metabolism of a wide range of endogenous and exogenous substances, including therapeutic drugs, with the result that an important mechanism for drug–alcohol interactions involves either an inhibition or induction of this enzyme. After taking a large acute dose of alcohol, the ethanol molecules compete with other drugs for detoxification enzymes. On the other hand, chronic consumption of large amounts of alcohol causes induction of the enzyme system so that alcoholics acquire an enhanced capacity for the metabolism of drugs via this system. Alcoholics are more prone to suffer from hepato-toxicity from taking acetaminophen (paracetamol) owing to their more active CYP2E1 detoxification enzymes, which convert acetaminophen into a chemically reactive and toxic intermediate compound, *N*-acetyl-*p*-benzoquinone imine (NAPQI).[27]

Chloral hydrate, one of the oldest sedative hypnotic drugs, has a complex interaction with ethanol. In the liver, chloral hydrate is rapidly reduced to its pharmacologically active metabolite 2,2,2-trichloroethanol, and this reaction is facilitated by the excess NADH (nicotinamide adenine dinucleotide) produced in the liver during the metabolism of ethanol. The pharmacologically active trichloroethanol is either conjugated with glucuronic acid and then excreted in the urine or is oxidized by ADH to an inactive metabolite, trichloroacetic acid. This latter ADH reaction is inhibited by the presence of ethanol, which competes for binding sites on the enzyme. The net result of both of these interactions with alcohol is that the effect of chloral hydrate on the individual is more intense and prolonged.[27]

Ethanol is a central nervous system depressant and a similar synergistic effect is found for other hypnotic drugs as well as antidepressants and narcotic analgesics so that allowance for ethanol–drug synergism is necessary when blood concentrations measured at autopsy are interpreted.[28] In general, any drug or chemical agent with psychoactive properties, that is, with its site of action in the central nervous system, has the potential for pharmacodynamic interaction with ethanol. The toxicity of the painkiller propoxyphene is enhanced in alcohol abusers and binge drinkers. Although the mechanism of the interaction with ethanol is not completely understood, it appears to be related to synaptic activity at GABA and opiate receptors with additive effects on the depression of respiration. Sedative-hypnotic drugs, which are widely prescribed, are obvious candidates for interaction with ethanol, owing to the similar mechanism and sites of action in the brain, namely, the GABA receptor complex. Examples of sedative-hypnotics exerting a pharmacodynamic interaction with ethanol include barbiturates and benzodiazepines. Deaths resulting from the combined use of large doses of alcohol and barbiturates are well recognized. However, in case of fatalities it has proved difficult to establish an added toxic effect from the concurrent use of diazepam with alcohol, so that in cases of acute alcohol poisoning the blood ethanol concentration is about the same, with or without the presence of diazepam.[29]

The alcohol flush reaction exhibited by many Japanese and other Asians when they drink alcoholic beverages is a direct consequence of acetaldehyde accumulation in the blood. Acetaldehyde reacts with β-adrenergic receptors of the autonomic nervous system leading to violent and unpleasant vasomotor responses. These individuals inherit a defective form of the enzyme acetaldehyde dehydrogenase (ALDH) and as a result are genetically prone to react strongly after drinking small amounts of alcohol, so that in effect they are equipped with a natural aversion to alcohol. A fatality has been reported in a Japanese subject with this inherited defect resulting in abnormally high concentrations of acetaldehyde after consumption of ethanol.[30] Disulfiram (tetraethylthiuram disulfide or Antabuse®) inhibits ALDH and is used in aversion therapy for treating alcoholics, although its clinical effectiveness has been debated. In a person taking the drug, subsequent ingestion of alcohol produces numerous unpleasant symptoms as a result of the toxic accumulation of acetaldehyde. Fatalities have been reported after reaching relatively low blood ethanol concentrations and with acetaldehyde in blood at concentrations between 12 and 41 mg/L.[31] When the edible inky cap mushroom (*Coprinus atramentarius*) is mistakenly eaten at the same time alcohol is consumed, the result is an Antabuse-like reaction owing to the presence of an unusual amino acid protoxin, coprine (*N*5-1-hydroxycyclopropyl-L-glutamine), which is broken down in the body to 1-aminocyclopropanol, an inhibitor of hepatic ALDH. Enzyme inhibition generally persists for about 72 h but may continue for 5 days, so consumption of ethanol during this period might produce an unpleasant Antabuse-like reaction. Metronidazole has been reported to cause death when taken with ethanol as a consequence of an Antabuse-like reaction,[32] but others have suggested that there might be an allergic reaction to the drug modified by the presence of alcohol.[33] Other prescription drugs that might produce an Antabuse-like reaction are chlorpropamide and tolbutamide, both oral hypoglycemic agents used to treat diabetes.

Cocaethylene is biosynthesized during the metabolism of cocaine in individuals with elevated BACs. This compound passes the blood–brain barrier to exert its pharmacological effects through the dopamine receptor thus enhancing the feelings of euphoria. Since the elimination half-life of cocaethylene is longer than for cocaine, it prolongs the effects of the drug on the individual. Several studies have suggested that when cocaine is taken together with ethanol, the risk of cardiotoxicity[34,35] and a fatal outcome[36] is increased. It has been suggested that ethanol enhances the acute toxicity of heroin and that ethanol use indirectly influences fatal heroin overdose through its association with infrequent (non-addictive) heroin use and thus a reduced tolerance to the acute toxic effects of heroin.[37] Although it has been proposed that carbon monoxide and ethanol may have an additive effect there is no conclusive evidence of this.

3.4 VITREOUS ALCOHOL

Analysis of vitreous humor (VH) is useful to corroborate a post-mortem BAC and this helps to distinguish ante-mortem intoxication from post-mortem synthesis of alcohol. VH can serve also as an alternative post-mortem specimen for alcohol analysis if for some reason blood is unavailable or contaminated. In most cases, the specimen of eye fluid is easily obtained and can be sampled without a full autopsy. Vitreous humor is a clear, serous fluid that is easy to work with analytically. Its anatomically isolated position protects it from bacterial putrefaction. In microbiological studies of vitreous obtained from 51 cadavers between 1 and 5 days post-mortem it was found that none of the samples contained large numbers of bacteria and only one contained fungi.[38]

The predictive value of a known vitreous humor alcohol concentration (VHAC) in estimating an unknown BAC in an individual case remains contentious, despite many studies.[39–50] Various formulae have been proposed, including a simple conversion factor, to predict BAC from VHAC but these do not take into account the uncertainty of the prediction for an individual subject. In one case series[51] simple linear regression was applied with BAC as outcome y variable and VHAC as predictor x variable (range 1 to 705 mg/dL). The regression equation was BAC = 3.03 + 0.852 VHAC with 95% prediction interval $\pm$ 0.019 $\sqrt{[7157272 + (\text{VHAC} - 189.7)^2]}$.

The practical application of the regression equation is shown in Table 3.1. Set out are the VHAC values that predict key BAC values of 80 and 150 mg/dL as the mean, or the minimum value at either the 95% or 99% prediction interval for the determination of a single BAC value. The prediction interval is too wide to be of much practical use. In addition, reanalysis of the raw data from previous publications gave significantly different regression equations in most instances. From an evidential viewpoint it would be unreasonable to give an estimate of the mean BAC based on the VHAC without also providing the 95% prediction interval, which is a measure of the degree of confidence attached to the estimate in an individual case.[52–54] Data from another large study suggest that dividing the VHAC by 2.0 would provide a very conservative estimate of the BAC, which is less than the true value with a high degree of confidence.[55]

Table 3.1 Prediction of Critical Values of BAC (as mean, or lower limit of 95% and 99% prediction intervals) from VHAC (mg/dL)[51]

Observed VHAC mg/dL	Predicted BAC mg/dL Mean	95% PI	99% PI
90	80	29–131	13–147
150	131	80–182	64–198
169	147	96–198	80–214
173	150	100–201	83–217
232	201	150–251	134–268
251	217	166–268	150–284

Note: PI = prediction interval for the determination of a single BAC value. Thus for an observed VHAC of 90 mg/dL the best estimate of BAC is 80 mg/dL; there is a 95% probability that the true value of BAC is between 29 and 131 mg/dL; and a 99% probability that the true value is between 13 and 147 mg/dL.

Blood has a lower water content than vitreous, which is approximately 98 to 99% water v/v, so the expectation is that the blood:vitreous alcohol ratio will be less than unity. In cases where the ratio of blood alcohol to vitreous humor alcohol concentration exceeds 1.0, the most likely explanation is that death occurred before diffusion equilibrium had been attained, which might have forensic significance.[41] Animal studies[56,57] indicate that, following intraperitoneal or intravenous injections of ethanol, BAC/VHAC ratios may be greater than 0.95 for 30 min or longer. A study of 43 fatalities[45] disclosed a bimodal distribution of blood:vitreous alcohol ratios with the first mode from 0.72 to 0.90 and a positively skewed distribution from 0.94 to 1.37. It seems that the first mode of the distribution ratio represents the elimination phase of the blood alcohol curve and that the second mode represents the absorption phase prior to equilibrium being established.

A second study of 86 cases confirmed this bimodal distribution and suggested that a blood/vitreous ratio greater than 0.95 indicates that death had occurred before equilibrium was achieved, and therefore in the early absorptive phase.[48] The blood/vitreous ratio during this early phase had a mean of 1.09 (SD = 0.38) in contrast to the late absorptive and elimination phases where the mean was 0.80 (SD = 0.09). Others[47] failed to reproduce this bimodal distribution. However, most deaths occur during the elimination phase, and it is clear that the observation of a bimodal distribution of blood/vitreous alcohol ratios in any study depends on the inclusion of cases dying in the absorptive phase. It is likely that the proportion of absorptive phase cases included in published series has varied considerably and that this accounts, at least in part, for the differences in published BAC/VHAC ratios.

It seems reasonable to assume that ethanol may diffuse into or out of the vitreous post-mortem. The chemical constituents of embalming fluid may diffuse into the vitreous humor after a body has been embalmed. Fortunately almost all commercial embalming fluids are free of ethyl alcohol, although they commonly contain methanol. A comparison of ethanol concentrations in 38 corpses both pre- and post-embalming suggested that there was no significant change in the vitreous humor ethanol concentration in the immediate aftermath of embalming.[58] However, in one case the embalmer cleaned the globus of the eye with ethanol on a cotton swab prior to placing an eye cap into position and this caused an elevation of vitreous ethanol from 0 to 340 mg/dL. In another fatality, an unusually high VHAC was attributed to prolonged post-mortem exposure of the eye surface to alcohol-containing vomit.[59] Conversely, prolonged submersion of a body in water may result in diffusion of alcohol out of the vitreous. This was the proposed explanation for finding a zero ethanol concentration in vitreous but concentrations of 370 mg/dL in urine and 223 mg/dL in blood (blood acetone 46 mg/dL) in a man submerged in cold fresh water for about 6 weeks. In a rabbit model that duplicated these circumstances the vitreous ethanol level fell from a mean of 196 to 30 mg/dL over the 6-week period.[60]

3.5 URINARY ALCOHOL

The ureteral urine, which is the urine as it is being formed, has a concentration of alcohol approximately 1.3 times that of whole blood. In fatalities the urine sample obtained is pooled bladder urine, which has accumulated over an unknown time interval between last urination and death. Consequently, the bladder urine alcohol concentration (UAC) does not necessarily reflect the BAC existing at the time of death. Instead it reflects the average BAC prevailing during the period that urine accumulates in the bladder since it was last emptied.

Several studies have examined the range of ratios between BAC and pooled bladder UAC at autopsy. One study[61] reported an average UAC/BAC ratio of 1.28:1 with a wide range of 0.21 to 2.66. In another study[44] the mean UAC/BAC ratio was 1.21:1 with a range from 0.22 to 2.07 using a direct injection gas chromatography (GC) technique and a mean ratio of 1.16:1 with a range of 0.20 to 2.10 with the more widely used headspace GC technique. In a large case series, simple linear regression analysis with BAC as outcome variable and autopsy bladder UAC as predictor variable (n = 435, range 3 to 587 mg/dL) gave the regression equation BAC = –5.6 + 0.811 UAC.[62] The 95% prediction interval around the regression line was given by the equation $\pm 0.026\sqrt{[9465804 + (\text{UAC} - 213.3)^2]}$. This shows that a BAC of 80 mg/dL was predicted with 95% certainty by a UAC of 204 mg/dL and similarly a BAC of 150 mg/dL by a UAC of 291 mg/dL. The prediction interval is very wide so that autopsy UAC has limited usefulness in predicting an unknown BAC for forensic purposes. Although an autopsy UAC should not be translated into a presumed BAC, it is possible to make a conservative estimate of the BAC existing during the time the urine was being produced and accumulated in the bladder, by dividing the observed autopsy UAC by 1.35 (or multiplying by three quarters). However, if the BAC profile were rising, which might be the case if death occurred soon after drinking ended, the calculation UAC/1.35 would underestimate the coexisting BAC.

When both urine and blood specimens are available at autopsy, then the UAC/BAC ratio may be of interpretive value, by giving an indication of the state of alcohol absorption and elimination in the body at the time of death.[63] A ratio less than unity or not more than 1.2 suggests, but does not prove, the existence of a rising BAC. If the UAC/BAC ratio exceeds 1.3, this suggests that the subject was in the post-absorptive stage at the time of death. The UAC/BAC ratio in 628 deaths from acute ethanol poisoning was 1.18 (SD = 0.182) and in 647 chronic alcoholic deaths was 1.30 (SD = 0.29), suggesting that the former group died typically before complete absorption and distribution of alcohol while the latter group died typically after complete absorption and distribution.[64] However, establishing whether a deceased had consumed alcohol immediately prior to death is most easily achieved by obtaining a sample of stomach contents at autopsy and analyzing this for alcohol. A gastric contents concentration less than 500 mg/dL has been taken to indicate a post-absorptive state.[43] Unusually high UAC/BAC ratios reflect urine accumulation over a long period of time and extreme ratios are well recognized as occurring in delayed deaths from acute alcohol poisoning.[65] In delayed traumatic deaths post-mortem urine ethanol concentrations may help establish or exclude the role of alcohol. Urine ethanol content of 200 mg/dL may be found with no alcohol present in the blood.[66]

3.6 ALTERNATIVE BIOLOGICAL SAMPLES

There have been many attempts to correlate post-mortem BACs with the concentrations of alcohol measured in a creative variety of specimens. In addition to the usual urine and vitreous humor these specimens have included saliva, cerebrospinal fluid, brain, liver, kidney, bone marrow, and skeletal muscle. All show a very wide range of variation in the ratio of the ethanol content in the target tissue or fluid compared with that in blood, making them of little value in practice. Since ethanol distribution in the body follows the distribution of body water, the ethanol content of these unconventional specimens is influenced by their water content. Cerebrospinal fluid is 98 to 99% water w/w and therefore generally will have a higher ethanol content than blood, whereas muscle, brain, liver, and kidney, with a water content of 75 to 78% w/w, will have lower concentrations of alcohol than blood. Moreover, the continuing post-mortem enzymatic activity within liver and kidney may reduce their ethanol content.

For blood alcohol levels greater than 40 mg/dL the average liver:heart blood ratio in 103 cases was 0.56, SD 0.30, with a range of 0 to 1.40.[67] However, liver is not recommended as a suitable sample for post-mortem ethanol analysis because it is rapidly invaded by gut microorganisms and provides abundant glycogen as substrate for ethanol production by fermentation, as well as being subject to post-mortem diffusion from gastric residue. For bile, in 89 cases with blood ethanol ranging from 46 to 697 mg/dL the bile:blood ratio averaged 0.99, range 0.48 to 2.04. Bile has been suggested as superior to vitreous humor in the estimation or corroboration of a blood alcohol level,[68] but, unfortunately, bile is vulnerable to post-mortem diffusion of unabsorbed alcohol in the stomach.[69] For cerebrospinal fluid (n = 54, blood ethanol range 46 to 697 mg/dL) the average ratio was 1.14 and range 0.79 to 1.64.[43]

Since ethanol impairs body functioning by its effects on the brain, it would seem logical to analyze brain tissue for alcohol at autopsy. This is not the practice for two reasons. First, the analysis of post-mortem blood is more helpful as it allows comparison with data available from living persons. Second, the brain is extremely heterogeneous so that within the brain the ethanol concentration may vary twofold to threefold between different regions; highest concentrations are found in the cerebellum and pituitary and lowest concentrations in the medulla and pons.[70] Brain tissue obtained from the frontal lobe (n = 33 blood ethanol range 72 to 388 mg/dL) gave an average brain:blood ratio of 0.86 with a range of 0.64 to 1.20.[43]

3.7 RESIDUAL GASTRIC ALCOHOL

Opinions vary as to how much alcohol may be found in the stomach post-mortem in case material. In one series[71] the highest concentration found was 2.95 g/dL and the author quoted a similar high of 5 g/dL in a previous study. In another series[72] only 1 of 60 cases had a concentration as high as 5.1 g/dL. In a small study[73] the highest concentration observed was 8.7 g/dL and this was in a suicidal hanging. Given that alcohol is rapidly absorbed from the stomach, it seems likely that death must occur within about an hour or less of ingesting substantial amounts of alcohol to detect a significant residue in the stomach post-mortem.

Researchers in the 1940s and 1950s debated the suitability of heart blood for quantitative analysis of ethanol on the grounds of possible artifactual elevation resulting from post-mortem diffusion of alcohol present in the stomach at the time of death.[74–80] Later investigations by Plueckhahn[71,81–86] led to the conclusion that post-mortem ethanol diffusion from the stomach into the pericardial sac and left pleural cavity was significant and could contaminate blood samples allowed to pool there, but cardiac chamber blood, as such, was not susceptible to this diffusion artifact to any significant extent. However, more recent case studies and a cadaver model have shown that cardiac chamber, aortic, and other torso blood samples may be significantly affected by gastric diffusion artifact.[73]

In one study[87] blood was obtained from the right atrium, ascending aorta, and the inferior vena cava in 307 subjects without significant decomposition and in whom the blood ethanol concentration was not less than 50 mg/dL in any sample. A total of 104 (33.9%) had one blood ethanol value 20% lower than the highest value. The most striking differences were found when gastric ethanol concentrations were greater than or equal to 800 mg/dL and with associated evidence of aspiration. In a second study[72] blood was obtained from the femoral vein, the aortic root, and the right atrium in 60 cases with blood ethanol concentrations of 50 mg/dL or greater and no gross trauma or significant decomposition. Although the mean alcohol concentrations for the different samples were not significantly different, there were wide variations in alcohol concentration among the various blood sample sites in a number of individual cases. Of the cases, 20 (33.3%) had within-case blood alcohol differences greater than 25%; 4 had differences greater than 50%, with 1 of these cases exceeding 400%. Indeed 3 of the 4 latter cases had gastric alcohol concentrations between 1 and 5.1 g/dL and the fourth had a concentration between 0.5 and 1 g/dL, whereas for the 60 cases as a whole, 22 were between 0.5 and 1 g/dL and 11 were above 1 g/dL. In a third study[73] of nine fatalities with known alcohol consumption shortly before death, two showed marked variations in blood ethanol concentrations in samples from ten sites, with ranges (mg/dL) of 97 to 238 and 278 to 1395; pericardial fluid 1060 and 686; vitreous humor 34 and 225; stomach contents 300 mL at 5.5 g/dL and 85 mL at 1.9 g/dL, respectively. These studies[72,73,87] suggest that post-mortem diffusion of alcohol from the stomach into the blood may be a significant, although uncommon problem.

The above findings were corroborated by a human cadaver model[73] with multiple blood site sampling after introducing 400 mL of alcohol solution (5%, 10%, 20%, or 40% weight/volume in water) into the stomach by esophageal tube. The pattern of ethanol diffusion showed marked between-case variability but typically concentrations were highest in pericardial fluid and, in decreasing order, in left pulmonary vein, aorta, left heart, pulmonary artery, superior vena cava, inferior vena cava, right heart, right pulmonary vein, and femoral vein. Diffusional flux was broadly proportional to the concentration of ethanol used, was time dependent (as assessed by 24-h and 48-h sampling), and was markedly inhibited by refrigeration at 4°C. After gastric instillation of 400 mL of 5% ethanol for 48 h at room temperature in paired cadavers, the concentrations (mg/dL) were as follows: pericardial fluid 135, 222; aorta 50, 68; left heart 77, 26; right heart 41, 28; femoral vein 0, 0. With a 10% solution of ethanol in the stomach, concentrations (mg/dL) were pericardial fluid 401, 255; aorta 129, 134; left heart 61, 93; right heart 31, 41; femoral vein 5, 7. The very high concentrations of alcohol found in the pericardial fluid emphasize the potential for serious contamination of any blood sample allowed to pool in the pericardial sac. Introducing 50 mL of 10% alcohol solution into the esophagus after esophago-gastric junction ligation produced similar aortic blood ethanol concentrations to those seen after gastric instillation. This suggests that post-mortem gastro-esophageal reflux and diffusion from the esophagus is one mechanism for artifactual elevation of aortic blood ethanol.

Post-mortem relaxation of the gastro-esophageal sphincter permits passive regurgitation of gastric contents into the esophagus, if body position and the volume of gastric contents permit. Thereafter, manipulation of the body during removal and transport might easily lead to contamination of the airways by gastric material, simulating agonal aspiration of vomitus. Alcohol in this gastric material could diffuse from the airways into the blood. An experimental study[88] demonstrated that a relatively small amount of ethanol introduced into the trachea of cadavers was readily absorbed into cardiac blood and also that there was direct diffusion from the trachea into both the aorta and superior vena cava.

3.8 POST-MORTEM STABILITY AND SYNTHESIS OF ALCOHOL

Alcohol loss from body tissues and fluids may occur post-mortem as a result of evaporation, shifts in the water content of tissues, enzymatic breakdown, and bacterial degradation. Following the death of an individual, many of the cells of the body survive for a variable time, providing the window of opportunity for tissue and organ transplantation from non-beating-heart donors. During this period of cellular life following somatic death, cellular enzymatic activity continues, ethanol is metabolized, and ethanol concentrations may fall slightly. Later bacterial invasion of tissues may result in further ethanol loss from bacterial degradation of ethanol. Ethanol is both utilized and produced by microorganisms, so that bodies with high initial levels may show a decrease, and bodies with low initial levels may show an increase.[89] However, in practice it is ethanol production by bacteria that represents the principal effect and this view is supported by an animal model.[90] Determining whether ethanol identified in post-mortem blood represents alcohol ingested prior to death or was formed post-mortem as a result of microbial activity is a common problem. Ethanol formation may occur in blood putrefying in a cadaver or in blood putrefying *in vitro*. It appears that ethanol is not formed post-mortem except by microbial action. Germ-free mice do not putrefy because of the absence of microorganisms,[91] and post-mortem autolysis of germ-free mice produces low levels of acetone and acetaldehyde but no ethanol. By contrast, putrefying conventional mice also produced ethanol, propionic acid, isopropyl alcohol, and *n*-propyl alcohol.[91]

Ethanol production in corpses[92] takes place by a pathway opposite to that of ethanol catabolism in the living body. The necessary alcohol dehydrogenase and acetaldehyde dehydrogenase enzymes are provided by the microorganisms associated with putrefaction while the carbohydrate substrates are present in blood and tissues. The level of tissue glycogen available for post-mortem glycolysis and subsequent microbial ethanol production varies considerably between tissues. Human liver contains about 1 to 8 g glycogen/100 g wet tissue, skeletal muscle 1 to 4 g/100 g, brain from a variety of animals 70 to 130 mg/100 g, and retina (ox) 90 mg/100 g (all figures calculated from dry weight assuming 75% water).[93] Anaerobic glycolysis produces pyruvate, the main substrate for ethanol production. Besides glucose, lactate is a source of pyruvate through the action of lactate dehydrogenase. Since lactate is found in relatively high concentrations in all post-mortem tissues (about 150 to 650 mg/100 g),[90] it may well be an important source of ethanol. A study *in vitro* with putrefying post-mortem blood under anaerobic conditions at room temperature demonstrated that ethanol formation occurred not only by way of glycolysis but also from lactate via conversion into pyruvate.[93] There is also evidence that ethanol might be produced by bacterial catabolism of amino acids and fatty acids.[94]

Escape of large numbers of bacteria from the gut occurs in the first instance via the lymphatics and portal venous system, within a few hours after death. At room temperature bacterial contamination of the systemic circulation occurs after about 6 h, and after 24 h there is direct bacterial penetration of the intestinal wall. Generally the tissues remain relatively free of viable bacteria during the first 24 h. Trauma immediately prior to death, intestinal lesions and neoplastic disease, generalized infection, and gangrene are all conditions associated with early spread of bacteria post-mortem.[90] In a living person, ethanol may be produced by gut bacteria present in the infected intraperitoneal fluid associated with a peritonitis following a stabbing.[95] A wide variety of bacteria normally present in the gut and responsible for putrefaction can generate ethyl alcohol in blood, brain, liver, and other tissues.[96] As well as bacteria, yeasts such as *Candida albicans* may be responsible for post-mortem alcohol production.[97] *Candida* overgrowth in the gut in the living may cause intra-gastrointestinal alcohol fermentation syndrome, a rare condition described in Japan.[98] *Candida* colonization of the gut, stagnation of food, and a high carbohydrate diet allow fermentation to produce sufficient alcohol to cause intoxication to the point of illness, with a BAC of 254 mg/dL reported in one case.[98]

There is considerable evidence that ethanol can be produced in corpses at levels up to 150 mg/dL after they have been stored for a few days at room temperature. In a study of 130 decomposing bodies,[99] there were 23 with presumed post-mortem ethanol production. Of these, 19 of 23 had blood ethanol concentrations of 70 mg/dL or less and the other 4 had levels of 110, 120, 130, and 220 mg/dL. Since both bacterial growth and enzyme activity are temperature dependent it is reasonable to assume that post-mortem ethanol production is inhibited by refrigeration. For example, a series of 26 in-hospital deaths refrigerated within 1 h of death and stored at 6°C for 3 to 27 h before autopsy showed no evidence of post-mortem ethanol production despite positive blood cultures in 13 cases, 7 of whom had blood glucose in excess of 20 mg/dL.[100] Chemical inhibitors of bacterial growth, such as sodium fluoride, prevent the production of ethanol in post-mortem specimens. As with all blood specimens intended for ethanol analysis it is recommended that post-mortem specimens should be preserved with 2% w/v sodium fluoride and that storage at temperatures above 4°C should be minimized.

Circumstances that can be expected to provide fertile ground for post-mortem ethanol production include prolonged exposure to a high environmental temperature, terminal hyperglycemia, and death from infectious disease with septicemia, natural disease such as ischemia affecting the large bowel, abdominal trauma, and severe trauma with wound contamination. Body disruption of a severity that commonly occurs in aviation accidents is associated with extensive microbial contamination and a resultant higher probability of post-mortem ethanol production. In a series of 975 victims of fatal aircraft accidents[101] the BAC exceeded 40 mg/dL in 79 cases and of these it was considered, based on ethanol distribution in urine, vitreous humor, and blood, that 27% represented post-mortem production and 28% ingestion, but 45% were unresolved. In such traumatic cases blood concentrations as high as 300 mg/dL might be synthesized post-mortem. In the *USS Iowa* explosion the highest ethanol concentration attributed to post-mortem fermentation was 190 mg/dL.[102]

In establishing whether there is post-mortem production or *in vivo* ingestion of alcohol, circumstantial evidence and corroborative analyses of vitreous humor and urine are of considerable assistance. Although vitreous contains glucose and lactate, both substrates for bacterial production of alcohol, it is protected by its remoteness from invasion by putrefactive bacteria spreading from the gut through the vascular system. Urine is useful because it normally contains little or no substrate for bacterial conversion to ethanol. An exception occurs if the urine contains sufficient suitable substrate, such as glucose, as a consequence of some pathological abnormality, particularly diabetes mellitus.[103]

After the early post-mortem period, when decomposition begins, the problem of production of ethanol increases because both decomposition and synthesis of ethanol are the result of microorganism spread and proliferation. This putrefactive phase starts about 2 or 3 days after death in a temperate climate but varies considerably depending on environmental conditions, primarily temperature. In early putrefaction when a sample of vitreous humor is still obtainable the presence of ethanol in this fluid is the best indicator of ethanol ingestion ante-mortem. The presence of ethanol in the urine, if it is available, is also a good indicator of ethanol ingestion. Once decomposition has progressed so that the vitreous is no longer available, due to collapse of the eyeballs, and blood cannot be obtained because the blood vessels are filled with putrefactive gases, then no reliance can be placed upon any sample, and interpretation of analytical results is hazardous if not impossible. In these cases anamnestic data may be more reliable than analytical data. In around 20% of decomposed bodies the ethanol detected is probably derived from endogenous sources based on its presence in blood but absence in vitreous or urine, but in many cases endogenous production cannot be distinguished from ingestion.[99] Post-mortem ethanol production in decomposed bodies results in blood or putrefactive fluid concentrations less than 150 mg/dL in over 95% of cases, but this general observation does not assist in evaluating a higher level found in an individual case.[99,104] In decomposing bodies endogenous ethanol production may occur in the bile as well as the blood and this might be particularly marked in the sanguineous putrefactive fluid, which accumulates in the pleural cavities.[104]

The importance of measuring ethanol concurrently in vitreous, urine, and blood samples was demonstrated in the assessment of low post-mortem BACs. A series of 381 cases with autopsy BACs less than 50 mg/dL was evaluated using the presence of ethanol in the vitreous and/or urine as indicators of ingestion rather than post-mortem production.[105] When the BAC was 10 to 19 mg/dL, the investigators found that 54% of cases had positive vitreous or urine ethanol concentrations (greater than 10 mg/dL); when the BAC was 20 to 29 mg/dL this percentage increased to 63%; when the BAC was 30 to 39 mg/dL the percentage was 73%; and when the BAC was 40 to 49 mg/dL, 92% of the cases had an alternative specimen positive for ethanol. Of the 165 cases where both vitreous and urine were available, more than 90% demonstrated consistent results; however, in 14 cases, there was an unexplained inconsistency with one specimen positive and the other negative.

Bacterial production of ethyl alcohol is associated with the production of other volatile compounds such as isoamyl alcohol, formaldehyde, n-propyl alcohol, isopropyl alcohol, propionic acid, acetone, acetic acid, acetaldehyde, n-butanol, sec-butanol, tert-butanol, n-butyric acid, and iso-butyric acid. Of these n-butyric acid and iso-butyric acid are said to be the most common associates of ethanol produced by putrefactive bacteria.[96] Others have advocated measuring n-propanol as a marker of microbial fermentation.[106] However, variability in metabolic pathways between micro-organisms leads to variability in the final products of glucose fermentation. Furthermore, there is no reliable quantitative relationship between these volatiles and ethanol produced post-mortem. This limits considerably the potential usefulness of measuring these other volatile products to distinguish between alcohol ingestion and post-mortem production.

A variety of biochemical approaches to the assessment of post-mortem synthesis of alcohol have been proposed. The major enzymatic pathway for disposal of ethanol is oxidative metabolism with Class I isozymes of ADH to produce acetaldehyde and this toxic metabolite is oxidized further by ALDH to acetic acid. Very small amounts of ingested ethanol (<1%) undergo non-oxidative metabolism by three different reaction mechanisms. Like many drug molecules with hydroxyl groups, ethanol undergoes a conjugation reaction catalyzed by liver enzymes to produce ethyl glucuronide (EtG) and ethyl sulfate (EtS).[107,108] Thus, EtG and EtS are specific metabolites of ethanol and these have proved very useful as markers of prior ethanol ingestion because elimination half-lives are much longer than for ethanol itself.[107,108] Analytical methods for ethyl glucuronide have been considerably improved in recent years such as by using liquid chromatography–mass spectrometry.[109]

Ethanol also reacts with a membrane-bound phospholipid to form phosphatidylethanol, and a third reaction is esterification with short-chain fatty acids to produce the corresponding esters.[110–112] These non-oxidative metabolites are measurable in blood and tissues and are excreted in urine with half-lives longer than ethanol itself.[112] Finding these non-oxidative metabolites, e.g., in urine taken at autopsy, is a strong indication that the deceased had consumed alcohol some time before death. The sensitivity of these markers compared with EtG and EtS remains to be established.

Another biochemical approach is based on the metabolic interaction between ethanol and the biogenic amine serotonin (Figure 3.1).[112] Without any consumption of alcohol, serotonin (5-hydroxytryptamine) undergoes oxidative deamination to 5-hydroxyindolealdehyde (5-HIALD).[112–114] This short-lived intermediate is rapidly oxidized to 5-hydroxyindoleacetic acid (5-HIAA) and a very small fraction (~1%) is reduced to 5-hydroxytryptophol (5-HTOL). Formation of the carboxylic acid (5-HIAA) is the dominant reaction and the 5HTOL/5-HIAA ratio in urine is normally very low (<15 nmol/μmol). However, after a person drinks alcohol the 5-HTOL/5-HIAA ratio increases appreciably owing to competitive inhibition for ALDH, which is now engaged in the metabolism of acetaldehyde derived from ethanol.[112–114] Measuring the 5-HTOL/5-HIAA ratio in urine has therefore found applications in clinical medicine to disclose consumption of alcohol, e.g., in alcohol and drug abusers, who must refrain from drinking as part of their treatment.[112] The ratio of serotonin metabolites can also be used to differentiate between ante-mortem ingestion of ethanol and post-mortem microbial synthesis.[113–115] Finding an elevated concentration of ethanol in urine but with a 5-HTOL/5-HIAA ratio below 15 nmol/μmol strongly suggests that microbial synthesis of ethanol has occurred.[115–117] However, the question of whether an elevated concentration of ethanol in a post-mortem specimen partly reflects microbial synthesis and partly ante-mortem ingestion cannot be resolved using these kinds of biochemical markers.

3.9 SEQUESTERED HEMATOMAS

That ethanol might be measured in sequestered hematomas was first suggested by Hirsch and Adelson,[118] although they claimed no originality, explaining that it is one of the "tricks of the trade." It has been most commonly applied to cases of head trauma with subdural or epidural hematomas,[118–121] but also to intracerebral clots.[122,123] Although primarily used for ethanol, any toxicant

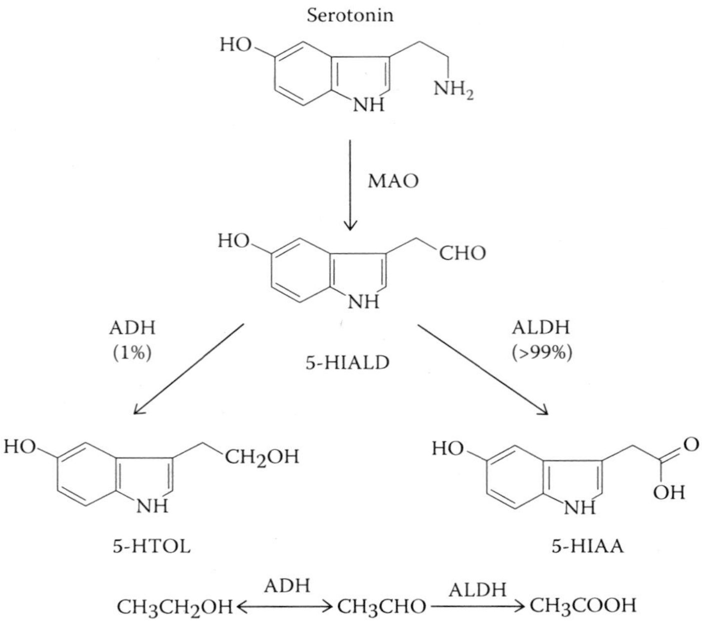

Serotonin

MAO

5-HIALD

ADH (1%)

ALDH (>99%)

5-HTOL

5-HIAA

$$CH_3CH_2OH \overset{ADH}{\longleftrightarrow} CH_3CHO \xrightarrow{ALDH} CH_3COOH$$

Figure 3.1 Serotonin and ethanol catabolism.

might be measured in the hematoma. From the accumulated case data (Table 3.2), it is clear that concentrations of ethanol in subdural hematomas may be markedly different from autopsy peripheral blood. In interpreting the significance of the results several possibilities should be considered. The hematoma may have developed rapidly at the time of injury, it may have been delayed and not developed for some hours, or it may have evolved over a period of time as the result of continuous or intermittent bleeding. If the hematoma accumulates over a period of hours, then its ethanol content will reflect a changing blood ethanol concentration during that time. Furthermore, the hematoma might not be perfectly sequestered and ethanol may diffuse both out of it and into it.

Table 3.2 Ethanol in Sequestered Hematomas

		Post-Mortem Ethanol Concentrations, mg/dL			
Ref.	**Survival Interval, h**	**Cardiac Blood**	**Peripheral Blood**	**Urine**	**SDH**
118	12	0		120	
	13	20		310	190
	6–12	160		310	300
	12–15	200			230 (R) 360 (L)
	>4	160		250	300
	UK	260		370	320
121	10	23	29	265	132
	1.5	121	121	232	206
	UK	58	47	226	104
	UK	151		322	192
119	13	0			120
	26	0			260
	9	40			100
	UK	50			150
	UK	40			1110

Note: UK = unknown; SDH = subdural hematoma.

Table 3.3 Ethanol in Ante-Mortem Blood and Sequestered Hematomas[120]

Time (h)		Ethanol (mg/dL)		
Injury to Pre-Mortem Sample	Pre-Mortem Sample to Death	Pre-Mortem	Post-Mortem	SDH
1	21	535	0	170
3	7	486	130	190
5	8.5	183	0	90
3	41	161	0	40
1 1/2	4.5	164	70	110
2	18	93	0	120
1	25.5	240	0	110
1	58	101	0	40

Note: SDH = subdural hematoma.

An animal model[121] has provided good experimental evidence that the current approach to determination of ethanol in sequestered hematomas is well founded.

In cases of head trauma associated with subdural or extradural hematomas and with a prolonged survival time, the autopsy blood ethanol concentration may be very low or even zero, whereas the ethanol concentration in the hematoma may be substantial, thus providing evidence that the deceased may have been intoxicated at the time of injury. In a study of 75 cases in which ethanol was measured in subdural hematomas and cardiac blood,[119] the analysis provided useful new information only in those cases with survival times greater than 9 h since it was these cases in which the blood ethanol had diminished markedly or been fully metabolized. In another case series consisting of 15 fatalities from penetrating and nonpenetrating head injuries,[120] there was a pre-mortem blood ethanol measurement available. Findings in nonpenetrating injuries (Table 3.3) and penetrating injuries were similar in that concentrations of ethanol in intracranial hematoma did not accurately reflect circulating blood concentrations at the time of injury. Therefore, quantitative interpretations must be cautious.

After suffering trauma, the development of an intracranial hemorrhage, either subdural or intracerebral, may be delayed. If the victim were intoxicated at the time of injury, then this delay may be sufficient to allow clearance of ethanol from the blood. The intracranial hematoma will then contain no ethanol despite the history of injury when intoxicated. This apparent conflict between the history of the circumstances of injury and the absence of ethanol in the hematoma has been used to provide corroboration that development of the hematoma had been delayed.[124]

3.10 ALCOHOLIC KETOACIDOSIS

Sudden death in a chronic alcoholic with a subsequent negative autopsy is a common problem. At autopsy such cases have only the stigmata of alcoholism, such as a fatty liver, and an inconsequential blood alcohol level. The mechanism behind these "fatty liver deaths" is obscure,[125] and recently the syndrome of alcoholic ketoacidosis (AKA) has been suggested as having a role.[126,127] The clinical literature on AKA is scant, and probably belies the true frequency of the syndrome,[128–135] but it does suggest that AKA is typically a relatively benign condition and only fatal when associated with some other disease process. The combination of starvation and alcohol abuse together precipitates the syndrome of AKA,[128] which is characterized by a metabolic acidosis, malnutrition, and binge drinking superimposed upon chronic alcohol abuse.[131] The typical clinical picture is that of an alcohol debauch terminated by anorexia with cessation of both food and alcohol intake, and finally, a variable period of hyperemesis. By the time AKA develops there is no measurable blood alcohol. The common symptoms of nausea, vomiting, and abdominal pain are accompanied by few objective physical findings and mental status is usually normal or only slightly impaired, but severe obtundation or coma occasionally occurs.[132]

Although the pathophysiology of AKA is complex, it seems that the pivotal variable is probably a relative deficiency of insulin. This results from starvation with consequent hepatic glycogen depletion, the inhibition of gluconeogenesis by an increased $NADH/NAD^+$ ratio resulting from metabolism of alcohol, and extracellular fluid volume depletion with α-adrenergic inhibition of insulin secretion.[133] Individuals with higher insulin levels would probably present with the syndrome of alcohol-induced hypoglycemia without ketoacidosis. Possibly the major factor separating AKA from alcohol-induced hypoglycemia is the dehydration and the starvation-induced α-adrenergic inhibition of insulin secretion in the former.

Extremely high free fatty acid levels in blood, ranging from 1800 to 3800 μEq/L, have been a consistent finding in clinical cases of AKA.[128,129] Beta-oxidation of these fatty acids generates acetyl-CoA, which is the precursor of ketones. Two critical enzymes in the ketogenesis pathway to acetoacetate are found in liver mitochondria only, [130] so that the liver is the only appreciable site of ketone body production. Excess acetoacetate in liver is converted into β-hydroxybutyrate and extrahepatic tissues can convert β-hydroxybutyrate back into acetoacetate and utilize both of these ketone bodies as respiratory substrates. Acetone is thought to be the product of a non-enzymatic decarboxylation of acetoacetate. Acetone production is therefore a function of the level of acetoacetate and the duration of its elevation, so that the presence of acetone is indicative of a sustained severe ketoacidosis.[134] However, the extent, if any, of post-mortem conversion of acetoacetate to acetone is not known and it is advisable and convenient to make a combined analysis of these two ketones in autopsy samples.[127] For convenience, the term "ketone bodies" is used to include only acetone, acetoacetate, and β-hydroxybutyrate. While this terminology is convenient because all three compounds are metabolically related, it is inaccurate because it excludes other biologically important ketones, e.g., pyruvate, but includes β-hydroxybutyrate, which does not have a ketone structure, although it is part of ketone metabolism.

The biochemical hallmark of AKA is ketoacidosis without marked hyperglycemia,[132] while by contrast diabetic ketoacidosis is defined by the triad of hyperglycemia, acidosis, and ketosis. In AKA the ratio of serum β-hydroxybutyrate to acetoacetate, which is normally 1:1, is increased to between 2:1 and 9:1, a ratio higher than that generally found in diabetic ketoacidosis.[135] The clinical diagnosis of AKA is hampered because the nitroprusside test (Acetest) for ketones in the urine is sensitive to acetoacetate but not to β-hydroxybutyrate.[129] In fasting subjects the concentration of acetoacetate in plasma might range up to 0.23 mmol/L, and that of β-hydroxybutyrate up to 0.65 mmol/L. In comparison, in AKA the corresponding concentrations of acetoacetate have range up to 7.5 mmol/L and of β-hydroxybutyrate up to 20.5 mmol/L.[127] In one autopsy study,[126] total ketones (mmol/L) in 71 non-alcoholics were as follows: vitreous 0.19 to 3.35, median 0.49; pericardial fluid 0.02 to 1.54, median 0.35; and femoral blood 0.23 to 8.08, median 1.00. The significantly high levels found in this autopsy series likely reflect deaths from chronic disease or prolonged agonal periods. Among 22 alcoholics, 18 had ketone levels not statistically different from non-alcoholics but there were 4 with femoral blood total ketone levels of 129.9 (also diabetic), 39.4 (no anatomical cause of death), 38.5 (suicidal hanging), and 18.6 (hypothermia), suggesting that while alcoholic ketoacidosis may be a previously overlooked potential cause of death, interpretation must be guarded and made within the total case context. This is in keeping with the clinical consensus that AKA is fatal only when associated with some other illness.

In the present state of knowledge it is uncertain whether AKA alone is sufficient to account for death, but it seems that very high levels of ketones (>10 mmol/L in femoral blood and >5 mmol/L in vitreous) are indicative of profound AKA, and these values can be expected in about 10% of all alcoholics subjected to a medicolegal autopsy. Both vitreous and pericardial fluid ketone levels are lower than those in post-mortem blood, possibly because they are less affected by any agonal or post-mortem changes. Alcoholic ketoacidosis can be diagnosed at autopsy by measurement of total ketone bodies (acetone, acetoacetate, and β-hydroxybutyrate) in vitreous humor, pericardial fluid, or peripheral blood. Finding significantly elevated levels is associated with a typical history of an alcoholic binge followed by a day or more of anorexia, and consequently an insignificant BAC. In alcoholics in whom the autopsy is negative (so-called fatty liver deaths), AKA may be an explanation for sudden death as a result of profound acidosis, with a critical fall in blood pH to around 7.0, precipitating vascular collapse.

3.11 POST-MORTEM MARKERS FOR ALCOHOL ABUSE

The prevalence of alcoholism in the forensic autopsy population varies between jurisdictions but can be as high as 10% or more. Poor hygiene and multiple bruises of different ages are more common in chronic alcoholics than in the general forensic autopsy population, and raise the index of suspicion in an individual case. The traditional method of diagnosing chronic alcoholism post-mortem is to evaluate the BAC and liver histology in the light of the available medical history. However, the presence of alcohol in the blood merely indicates alcohol ingestion prior to death and almost half of all alcoholics die with a zero BAC. Also, the pathological features of alcoholic liver disease are relatively nonspecific and the extent of liver disease in many alcoholics is no worse than in the general forensic autopsy population.

The spectrum of alcoholic liver disease includes hepatomegaly, steatosis with or without lipogranulomas, alcoholic hepatitis, and cirrhosis.[136] This complete spectrum, including alcoholic-type hepatitis, can be perfectly mimicked by a few non-alcohol-related conditions such as obesity with or without dieting, jejuno-ileal bypass for obesity, diabetes mellitus, and perhexilline maleate toxicity. Hepatic steatosis is the most common form of alcoholic liver disease seen at necropsy, and significant steatosis may be induced by the amounts of alcohol consumed by many social drinkers. Following the withdrawal of alcohol, the mobilization of this accumulated fat begins in 1 to 2 days and is complete in 4 to 6 weeks even in severe cases. While hepatic steatosis is potentially reversible, alcoholic hepatitis is thought by some to represent the point of no return within the spectrum, for once this stage is reached the disorder tends to progress to cirrhosis. Even so, a person with alcoholic hepatitis may be symptom-free and performing normal social functions. In general, there is not always a good correlation between symptoms and morphological findings in alcoholic hepatitis. Histologically, alcoholic hepatitis is characterized by liver cell necrosis with a predominantly neutrophil polymorph reaction and peri-cellular fibrosis. The hepatitis is mainly centrilobular in distribution and classically associated with the presence of Mallory's hyaline. More than half of all cases of cirrhosis coming to autopsy are the result of alcohol abuse. The classical alcoholic cirrhosis is micronodular and fatty, but this is not necessarily so since it may not be fatty and may evolve into a macronodular cirrhosis. Persons abusing alcohol frequently abuse other drugs and may develop drug-related hepatotoxicity, in particular late acetaminophen (paracetamol) toxicity from excessive, but not suicidal, doses of this antipyretic drug.

The search for a corroborative post-mortem biochemical marker for chronic alcoholism has taken as a starting point those clinical studies on the value of biochemical markers of alcoholism in the living. The serum enzyme γ-glutamyltransferase (GGT) is one of the most frequently used clinical markers and has a reported sensitivity of 39 to 87% but a specificity of only 11 to 50%.[137] This poor specificity is mainly the result of interference by various hepatic and other diseases, and drug therapy. At autopsy the difficulties are greater because GGT is subject to significant post-mortem changes. GGT levels in right heart blood may be two to eight times greater than in femoral venous blood owing to post-mortem diffusion of GGT from the liver. Furthermore, post-mortem hemolysis interferes with some quantitative enzymatic GGT methods. More recently, carbohydrate-deficient transferrin (CDT) has been used as a clinical marker of alcoholism, offering 83 to 90% sensitivity and 99% specificity.[137] CDT is thought to become elevated at the threshold of hazardous drinking, which is generally accepted as being 50 to 80 g/day. An assessment of the value of CDT in diagnosing chronic alcoholism at autopsy concluded that it might have a sensitivity of 70% and a specificity of 85% if the cut-off value for the diagnosis of alcohol abuse post-mortem were raised above the accepted clinical cut-off value.[138] This suggests that both CDT and GGT are likely to be subject to post-mortem changes.

Trace amounts of methanol (less than 1.0 mg/L) are produced in the body in the course of intermediary metabolism and the endogenous levels increase during a period of heavy drinking. Ingestion of methanol as a congener in various alcohol beverages adds to this accumulation.[139] When alcoholics consume alcohol over a period of several days or weeks reaching blood ethanol concentrations of 150 to 450 mg/dL, then the methanol levels in blood and urine progressively increase to 20 to 40 mg/L. The elimination of methanol lags behind ethanol by 12 to 24 h and follows approximately the same time course as ethanol withdrawal symptoms leading to speculation on the role of methanol and/or its metabolites in alcohol withdrawal and hangover.[139] Below a blood ethanol concentration of about 10 mg/dL, hepatic ADH is no longer saturated with its preferred substrate and the metabolism of methanol can therefore commence. At this low concentration the elimination of ethanol follows first-order kinetics with a half life of 15 min.[139] The half-life of methanol, however, is about ten times longer. As a result elevated concentrations of methanol will persist in blood for about 10 h after ethanol has reached endogenous levels and can serve as a marker of recent heavy drinking.[139]

Blood methanol levels in 24 teetotalers ranged from 0.1 to 0.8 mg/L with a mean of 0.44 mg/L so that these levels can be regarded as physiological. By contrast, blood methanol concentrations in samples taken on admission of 20 chronic alcoholics to hospital ranged from 0.22 to 20.1 mg/L.[140] The general extent to which methanol may accumulate in the blood of chronic alcoholics can be gauged from a study of ethanol and methanol in blood samples from 519 drunk-driving suspects.[139] The concentration of ethanol ranged from 0.01 to 3.52 mg/g and the concentration of methanol in the same sample ranged from 1 to 23 mg/L with a mean of 7.3 (SD 3.6) and a positively skewed distribution. By contrast, in 15 fatalities following hospital admission for methanol poisoning the concentrations of methanol in post-mortem blood from the heart ranged from 23 to 268 mg/dL.[141]

3.12 METHANOL

Methanol (wood alcohol) is used as antifreeze, photocopier developer, a paint remover, a solvent in varnishes, and a denaturant of ethanol, and is readily available as methylated spirit. It may be used also as a substitute for ethanol by alcoholics.[142] The distribution of methanol in body fluids (including vitreous humor) and tissues was reported as similar to that of ethanol, but there may be preferential concentration in liver and kidney.[143,144] The lethal dose of methanol in humans shows pronounced individual differences ranging from 15 to 500 mL. Clusters of poisonings are seen secondary to consumption of adulterated beverages.[141,145,146]

Acute methanol poisoning produces a distinct clinical picture with a latent period of several hours to days between consumption and the appearance of first symptoms. A combination of blurred vision with abdominal pain and vomiting is found in the majority of victims within the first 24 h after presentation. Visual disturbances, pancreatitis, metabolic acidosis, and diffuse encephalopathy may be seen in severe cases.[145] The characteristic delay between ingestion and onset of symptoms is thought to reflect the delayed appearance of metabolites (formaldehyde and formic acid), which are more toxic than methanol itself.

Methanol poisoning is characterized by a metabolic acidosis with an elevated anion gap. The serum anion gap is defined as (sodium + potassium) to (bicarbonate + chloride), and represents the difference in unmeasured cations and unmeasured anions, which includes organic acids. Both formic acid, produced by methanol catabolism, and lactic acid, resulting from disturbed cellular metabolism, are responsible for the metabolic acidosis.[147] The severity of the poisoning correlates with the degree of metabolic acidosis more closely than with the blood concentration of methanol.[148] Measuring formic acid concentrations may be of some value in assessing methanol poisoning. Reported formic acid levels in two methanol fatalities were 32 and 23 mg/dL in blood and 227 and 47 mg/dL in urine.[149] One well-documented outbreak of methanol poisoning[150] involved 59 people, 8 of whom died outside hospital while 51 were hospitalized, and of these a further 9 died in hospital, 5 survived with sequelae, and 1 died a year later of cerebral sequelae. In the 51 hospitalized victims, who had a median age of 53 years, the serum concentration of methanol (range 10 to 470 mg/dL, median 80 mg/dL) proved a poor predictor of survival or visual sequelae. Respiratory arrest or coma on hospital admission was associated with 75 and 67% mortality, respectively. Overall, prognosis was closely correlated with the degree of metabolic acidosis, so that a fatal outcome was associated with a pH < 6.9 and base deficit > 28 mmol/L, with an inadequate ability to compensate for the metabolic acidosis by hyperventilation being reflected in an increased blood pCO_2.

Methanol poisoning has a high mortality mainly because of delay in diagnosis and treatment.[150] The standard treatment includes the competitive inhibition of methanol oxidation by the intravenous administration of ethanol, thus preventing the formation of toxic metabolites, formaldehyde and formic acid. Both methanol and ethanol are substrates for hepatic ADH, although the affinity of the enzyme is much higher for ethanol than for methanol by about 10:1.[151] Consequently, the biotransformation of methanol into its toxic metabolites can be blocked by the administration of ethanol.[152] One disadvantage of ethanol is that it exerts its own depressant effect on the central nervous system at the steady-state concentration of 100 to 120 mg/dL in blood, which must be maintained for many hours.[152,153] A more modern, and expensive, antidote for methanol poisoning is fomepizole (4-methyl pyrazole or Antizol®), which also acts as a competitive inhibitor of alcohol dehydrogenase.[154,155] This drug is preferred to ethanol for treating children and adults with known liver dysfunction.[156] During fomepizole treatment, the concentration of formate in blood may be a better prognostic indicator than the methanol concentration.[157]

3.13 ISOPROPYL ALCOHOL

Isopropyl alcohol (isopropanol) is used as a substitute for ethanol in many industrial processes and in home cleaning products, antifreeze, and skin lotions. A 70% solution is sold as "rubbing alcohol" and may be applied to the skin and then allowed to evaporate, as a means of reducing body temperature in a person with fever. Isopropanol has a characteristic odor and a slightly bitter taste. Although much less dangerous than methanol, deaths have been reported following accidental ingestion of isopropanol, e.g., in alcoholics who use it as an ethanol substitute.[158] Fatalities may occur rapidly as a result of central nervous system depression or may be delayed, when the presence or absence of shock with hypotension is the most important single prognostic factor.

Isopropyl alcohol has an apparent volume of distribution of 0.6 to 0.7 L/kg, being similar to that of ethanol and with distribution complete within about 2 h.[159,160] Elimination most closely approximates first-order kinetics although this is not well defined ($t_{1/2}$ = 4 to 6 h).[161] This secondary alcohol is metabolized to acetone, predominantly by liver alcohol dehydrogenase, and approximately 80% is excreted as acetone in the urine with 20% excreted unchanged.[161,162] The acetone causes a sweet ketonic odor on the breath. The elimination of both isopropanol and its major metabolite acetone obeyed apparent first-order kinetics with half-lives of 6.4 and 22.4 h, respectively, in a 46-year-old non-alcoholic female with initial serum isopropanol and acetone concentrations of 200 and 12 mg/dL, respectively.[161]

In a review of isopropanol deaths, 31 were attributed to isopropanol poisoning alone, and the blood isopropanol concentrations ranged from 10 to 250 mg/dL, mean 140 mg/dL, and acetone ranged from 40 to 300 mg/dL, mean 170 mg/dL.[163] Four cases with low blood isopropanol levels (10 to 30 mg/dL) had very high acetone levels (110 to 200 mg/dL). For this reason both acetone and isopropanol should be measured in suspected cases of isopropanol poisoning.

High blood levels of acetone may be found in diabetes mellitus and starvation ketosis, which opens the possibility that ADH might reduce acetone to isopropyl alcohol. This is the suggested explanation for the detection of isopropyl alcohol in the blood of persons not thought to have ingested this compound. In 27 such fatalities blood isopropyl alcohol ranged from less than 10 to 44 mg/dL with a mean of 14 mg/dL, and in only 3 cases was the concentration greater than 20 mg/dL. Acetone levels ranged up to 56 mg/dL and in no individual case did the combined isopropanol and acetone levels come close to those seen in fatal isopropyl alcohol poisoning.[164]

3.14 CONCLUDING REMARKS

Blood ethanol concentration can be expected to be positive in around one half of all unnatural deaths so that routine screening of such deaths for ethanol is highly desirable. For natural deaths as a whole, the return of positives is not sufficiently high to justify screening, unless there is a history of chronic alcoholism or of recent alcohol ingestion. The autopsy blood sample should never be obtained from the heart, aorta, or other large vessels of the chest or abdomen or from blood permitted to pool at autopsy in the pericardial sac, pleural cavities, or abdominal cavity. If by mischance such a specimen is the only one available, then its provenance should be clearly declared and taken into account in the interpretation of the analytical results. Blind needle puncture of the chest to obtain a "cardiac" blood sample or a so-called "subclavian stab" is not recommended because at best it produces a chest cavity blood sample of unknown origin and at worst a contaminated sample. The most appropriate routine autopsy blood sample for ethanol analyses, as well as other drug analyses, is one obtained from either the femoral vein or the external iliac vein using a needle and syringe after clamping or tying off the vessel proximally. The sample should be obtained early in the autopsy and prior to evisceration. Samples of vitreous humor and urine, if the latter is available, should also be taken. The interpretation of the significance of the analytical results of these specimens must, of necessity, take into account the autopsy findings, circumstances of death, and recent history of the decedent. To attempt to interpret the significance of an alcohol level in an isolated autopsy blood sample without additional information is to invite a medicolegal disaster.

REFERENCES

1. Senkowski, C.M., Senkowski, B.S., and Thompson, K.A., The accuracy of blood alcohol analysis using headspace gas chromatography when performed on clotted samples, *J. Forensic Sci.,* 35, 176, 1990.
2. Hodgson, B.T. and Shajani, N.K., Distribution of ethanol: plasma to the whole blood ratios, *Can. Soc. Forensic Sci. J.,* 18, 73, 1985.
3. Hak, E.A., Gerlitz, B.J., Demont, P.M., and Bowthorpe, W.D., Determination of serum alcohol: blood alcohol ratios, *Can. Soc. Forensic Sci. J.,* 28, 123, 1995.
4. Felby, S. and Nielsen, E., The postmortem blood alcohol concentration and the water content, *Blutalkohol,* 31, 24, 1994.
5. Wallgren, H. and Barry, H., *Actions of Alcohol,* Elsevier, Amsterdam, 1970.
6. Johnson, H.R.M., At what blood levels does alcohol kill, *Med. Sci. Law,* 25, 127, 1985.
7. Jones, A.W. and Holmgren, P., Comparison of blood-ethanol concentration in deaths attributed to acute alcohol poisoning and chronic alcoholism, *J. Forensic Sci.*, 48, 874, 2003.
8. Niyogi, S.K., Drug levels in cases of poisoning, *Forensic Sci.,* 2, 67, 1973.
9. Taylor, H.L. and Hudson, R.P.J., Acute ethanol poisoning: a two-year study of deaths in North Carolina, *J. Forensic Sci.,* 639, 1977.
10. Wilson, C.I., Ignacio, S.S., and Wilson, G.A., An unusual form of fatal ethanol intoxication, *J. Forensic Sci.,* 50, 676, 2005.
11. Padosch, S.A., Schmidt, P.H., Kroner, L.U., and Madea, B., Death due to positional asphyxia under severe alcoholisation: pathophysiologic and forensic considerations, *Forensic Sci. Int.*, 149, 67, 2005.
12. Bell, M.D., Rao, V.J., Wetli, C.V., and Rodriguez, R.N., Positional asphyxiation in adults, *Am. J. Forensic Med. Pathol.*, 13, 101, 1992.

13. Perper, J.A., Twerski, A., and Wienand, J.W., Tolerance at high blood alcohol concentrations: a study of 110 cases and review of the literature, *J. Forensic Sci.,* 31, 212, 1986.
14. Davis, A.R. and Lipson, A.H., Central nervous system tolerance to high blood alcohol levels, *Med. J. Aust.,* 144, 9, 1986.
15. Lindblad, B. and Olsson, R., Unusually high levels of blood alcohol? *J. Am. Med. Assoc.,* 236, 1600, 1976.
16. Johnson, R.A., Noll, E.C., and MacMillan Rodney, W., Survival after a serum ethanol concentration of 1 1/2%, *Lancet,* 1394, 1982.
17. Poklis, A. and Pearson, M.A., An unusually high blood ethanol level in a living patient, *Clin. Toxicol.,* 10, 429, 1977.
18. Hammond, K.B., Rumack, B.H., and Rodgerson, D.O., Blood ethanol: a report of unusually high levels in a living patient, *J. Am. Med. Assoc.,* 226, 63, 1973.
19. Berild, D. and Hasselbalch, H., Survival after a blood alcohol of 1127 mg/dL, *Lancet,* 383, 1981.
20. O'Neill, S., Tipton, K.F., Prichard, J.S., and Quinlan, A., Survival after high blood alcohol levels: association with first-order elimination kinetics, *Arch. Intern. Med.,* 144, 641, 1984.
21. Jones, A.W., The drunkest drinking driver in Sweden: blood-alcohol concentration 545%, *J. Stud. Alcohol.,* 60, 400, 1999.
22. Kortelainen, M., Drugs and alcohol in hypothermia and hyperthermia related deaths: a respective study, *J. Forensic Sci.,* 32, 1704, 1987.
23. Teige, B. and Fleischer, E., Blodkonsentrasjoner ved akutte forgiftningsdodsfall, *Tidsskr. Nor. Laegeforen.,* 103, 679, 1983.
24. Teresinki, G., Buszewicz, G., and Madro, R., Biochemical background of ethanol-induced cold susceptibility, *Legal Med.,* 7, 15, 2005.
25. Kortelainen, M., Hyperthermia deaths in Finland in 1970–86, *Am. J. Forensic Med. Pathol.,* 12, 115, 1991.
26. Ramsay, D.A. and Shkrum, M.J., Homicidal blunt head trauma, diffuse axonal injury, alcoholic intoxication, and cardiorespiratory arrest: a case report of a forensic syndrome of acute brainstem dysfunction, *Am. J. Forensic Med. Pathol.,* 16, 107, 1995.
27. Jones, A.W., Alcohol and drug interactions, in *Handbook of Drug Interactions*, Mozayani, A. and Raymon, L.P., Eds., Humana Press, Totowa, NJ, 2003, 395.
28. King, L.A., Effect of ethanol on drug levels in blood in fatal cases, *Med. Sci. Law,* 22, 233, 1982.
29. Holmgren, P. and Jones, A.W., Coexistence and concentrations of ethanol and diazepam in post-mortem blood specimens: risk for enhanced toxicity? *J. Forensic Sci.,* 48, 1416, 2003.
30. Yamamoto, H., Tangegashima, A., Hosoe, H., and Fukunaga, T., Fatal acute alcohol intoxication in an ALDH2 heterozygote, *Forensic Sci. Int.,* 112, 201, 2000.
31. Heath, M.J., Pachar, J.V., Perez Martinez, A.L., and Toseland, P.A., An exceptional case of lethal disulfiram-alcohol reaction, *Forensic Sci. Int.,* 56, 45, 1992.
32. Cina, S.J., Russell, R.A., and Conradi, S.E., Sudden death due to metronidazole-ethanol interaction, *Am. J. Forensic Med. Pathol.,* 17, 343, 1996.
33. Williams, C.S. and Woodcock, K.R., Do ethanol and metronidazole interact to produce a disufiram-like reaction? *Ann. Pharmacother.,* 34, 255, 2000.
34. Jatlow, P., Cocaethylene. Pharmacologic activity and clinical significance, *Ther. Drug Monitor.,* 15, 533, 1993.
35. Jatlow, P., McChance, E.F., Bradberry, C.W., Elsworth, J.D., Taylor, J.R., and Roth, R.H., Alcohol plus cocaine: the whole is more than the sum of its parts, *Ther. Drug Monitor.,* 18, 460, 1996.
36. Tardiff, K., Marzuk, M.P., Leon, A.C., Hirsch, C.S., Stajic, M., Portera, L., and Hartwell, N., Cocaine, opiates, and ethanol in homicides in New York City: 1990 and 1991, *J. Forensic Sci.,* 40, 387, 1995.
37. Ruttenber, A.J., Kalter, H.D., and Santinga, P., The role of ethanol abuse in the etiology of heroin-related death, *J. Forensic Sci.,* 35, 891, 1990.
38. Harper, D.R., A comparative study of the microbiological contamination of postmortem blood and vitreous humour samples taken for ethanol determination, *Forensic Sci. Int.,* 43, 37, 1989.
39. Sturner, W.Q. and Coumbis, R.J., The quantitation of ethyl alcohol in vitreous humor and blood by gas chromatography, *Am. J. Clin. Pathol.,* 46, 349, 1966.
40. Leahy, M.S., Farber, E.R., and Meadows, T.R., Quantitation of ethyl alcohol in the postmortem vitreous humor, *J. Forensic Sci.,* 13, 498, 1968.

41. Felby, S. and Olsen, J., Comparative studies of postmortem ethyl alcohol in vitreous humor, blood, and muscle, *J. Forensic Sci.,* 14, 93, 1969.
42. Coe, J.I. and Sherman, R.E., Comparative study of postmortem vitreous humor and blood alcohol, *J. Forensic Sci.,* 15, 185, 1970.
43. Backer, R.C., Pisano, R.V., and Sopher, I.M., The comparison of alcohol concentrations in postmortem fluids and tissues, *J. Forensic Sci.,* 25, 327, 1996.
44. Winek, C.L. and Esposito, F.M., Comparative study of ethanol levels in blood versus bone marrow, vitreous humor, bile and urine, *Forensic Sci. Int.,* 17, 27, 1981.
45. Caughlin, J.D., Correlation of postmortem blood and vitreous humor alcohol concentration, *Can. Soc. Forensic Sci. J.,* 16, 61, 1983.
46. Stone, B.E. and Rooney, P.E., A study using body fluids to determine blood alcohol, *J. Anal. Toxicol.,* 8, 95, 1984.
47. Jollymore, B.D., Fraser, A.D., Moss, M.A., and Perry, R.A., Comparative study of ethyl alcohol in blood and vitreous humor, *Can. Soc. Forensic Sci. J.,* 17, 50, 1984.
48. Yip, D.C. and Shum, B.S., A study on the correlation of blood and vitreous humour alcohol levels in the late absorption and elimination phases, *Med. Sci. Law,* 30, 29, 1990.
49. Caplan, Y.H. and Levine, B., Vitreous humor in the evaluation of postmortem blood ethanol concentrations, *J. Anal. Toxicol.,* 14, 305, 1990.
50. Neil, P., Mills, A.J., and Prabhakaran, V.M., Evaluation of vitreous humor and urine alcohol levels as indices of blood alcohol levels in 75 autopsy cases, *Can. Soc. Forensic Sci. J.,* 18, 97, 1985.
51. Pounder, D.J. and Kuroda, N., Vitreous alcohol is of limited value in predicting blood alcohol, *Forensic Sci. Int.,* 65, 73, 1994.
52. Pounder, D.J. and Kuroda, N., Vitreous alcohol: the author's reply, *Forensic Sci. Int.,* 73, 159, 1995.
53. Kraut, A., Vitreous alcohol, *Forensic Sci. Int.,* 73, 157, 1995.
54. Yip, D.C.P., Vitreous humor alcohol, *Forensic Sci. Int.,* 73, 155, 1995.
55. Jones, A.W. and Holmgren, P., Uncertainty in estimating blood ethanol concentrations by analysis of vitreous humour, *J. Clin. Pathol.*, 54, 699, 2001.
56. Olsen, J.E., Penetration rate of alcohol into the vitreous humor studied with a new in vivo technique, *Acta Ophthalmol. Copenh.,* 49, 585, 1971.
57. Fernandez, P., Lopez-Rivadulla, M., Linares, J.M., Tato, F., and Bermejo, A.M., A comparative pharmacokinetic study of ethanol in the blood, vitreous humour and aqueous humour of rabbits, *Forensic Sci. Int.,* 41, 61, 1989.
58. Scott, W., Root, R., and Sanborn, B., The use of vitreous humor for determination of ethyl alcohol in previously embalmed bodies, *J. Forensic Sci.,* 913, 1974.
59. Singer, P.P. and Jones, G.R., Very unusual ethanol distribution in a fatality, *J. Anal. Toxico*l., 21, 506, 1997.
60. Basu, P.K., Avaria, M., Jankie, R., Kapur, B.M., and Lucas, D.M., Effect of prolonged immersion on the ethanol concentration of vitreous humor, *Can. Soc. Forensic Sci. J.,* 16, 78, 1983.
61. Kaye, S. and Cardona, E., Errors of converting a urine alcohol value into a blood alcohol level, *Am. J. Clin. Pathol.,* 52, 577, 1969.
62. Kuroda, N., Williams, K., and Pounder, D.J., Estimating blood alcohol from urinary alcohol at autopsy, *Am. J. Forensic Med. Pathol.,* 16, 219, 1995.
63. Levine, B. and Smialek, J.E., Status of alcohol absorption in drinking drivers killed in traffic accidents, *J. Forensic Sci.,* 45, 3, 2000.
64. Jones, A.W. and Holmgren, P., Urine/blood ratios of ethanol in deaths attributed to acute alcohol poisoning and chronic alcoholism, *Forensic Sci. Int.*, 135, 206, 2003.
65. Kaye, S. and Hag, H.B., Terminal blood alcohol concentrations in ninety-four fatal cases of acute alcoholism, *J. Am. Med. Assoc.,* 451, 1957.
66. Alha, A.R. and Tamminen, V., Fatal cases with an elevated urine alcohol but without alcohol in the blood, *J. Forensic Med.,* 11, 3, 1964.
67. Jenkins, A.J., Levine, B.S., and Smialek, J.E., Distribution of ethanol in postmortem liver, *J. Forensic Sci.,* 40, 611, 1995.
68. Stone, B.E. and Rooney, P.A., A study using body fluids to determine blood alcohol, *J. Anal. Toxicol.,* 8, 95, 1984.
69. Pounder, D.J., Fuke, C., Cox, D.E., Smith, D., and Kuroda, N., Postmortem diffusion of drugs from gastric residue, *Am. J. Forensic Med. Pathol.,* 17, 1, 1996.

70. Christopoulos, G., Kirch, E.R., and Gearien, J.E., Determination of ethanol in fresh and putrefied postmortem tissues, *J. Chromatogr.,* 87, 455, 1973.
71. Plueckhahn, V.D., Alcohol levels in autopsy heart blood, *J. Forensic Med.,* 15, 12, 1968.
72. Briglia, E.J., Bidanset, J.H., Dal Cortivo, L.A., The distribution of ethanol in postmortem blood specimens, *J. Forensic Sci.,* 37, 991, 1992.
73. Pounder, D.J. and Smith, D.R.W., Postmortem diffusion of alcohol from the stomach, *Am. J. Forensic Med. Pathol.,* 16, 89, 1995.
74. Bowden, K.M. and McCallum, N.E.W., Blood alcohol content: some aspects of its post-mortem uses, *Med. J. Aust.,* 2, 76, 1949.
75. Gifford, H. and Turkel, H.W., Diffusion of alcohol through stomach wall after death: a cause of erroneous postmortem blood alcohol levels, *J. Am. Med. Assoc.,* 161, 866, 1956.
76. Turkel, H.W. and Gifford, H., Erroneous blood alcohol findings at autopsy; avoidance by proper sampling technique, *J. Am. Med. Assoc.,* 164, 1077, 1957.
77. Turkel, H.W. and Gifford, H., Blood alcohol [letter], *J. Am. Med. Assoc.,* 165, 1993, 1957.
78. Heise, H.A., Erroneous postmortem blood alcohol levels [letter], *J. Am. Med. Assoc.,* 165, 1739, 1957.
79. Muehlberger, C.W., Blood alcohol findings at autopsy [letter], *J. Am. Med. Assoc.,* 165, 726, 1957.
80. Harger, R.N., Heart blood vs. femoral vein blood for postmortem alcohol determinations [letter], *J. Am. Med. Assoc.,* 165, 725, 1957.
81. Plueckhahn, V.D., The significance of blood alcohol levels at autopsy, *Med. J. Aust.,* 15, 118, 1967.
82. Plueckhahn, V.D. and Ballard, B., Factors influencing the significance of alcohol concentrations in autopsy blood samples, *Med. J. Aust.,* 1, 939, 1968.
83. Plueckhahn, V.D., The evaluation of autopsy blood alcohol levels, *Med. Sci. Law,* 8, 168, 1968.
84. Plueckhahn, V.D. and Ballard, B., Diffusion of stomach alcohol and heart blood alcohol concentration at autopsy, *J. Forensic Sci.,* 12, 463, 1967.
85. Plueckhahn, V.D., Postmortem blood chemistry — the evaluation of alcohol (ethanol) in the blood, in *Recent Advances in Forensic Pathology,* Camps, F.E., Ed., Churchill, London, 1969, 197.
86. Plueckhahn, V.D., The significance of alcohol and sugar determinations in autopsy blood, *Med. J. Aust.,* 10, 46, 1970.
87. Marraccini, J.V., Carroll, T., Grant, S., Halleran, S., and Benz, J.A., Differences between multisite postmortem ethanol concentrations as related to agonal events, *J. Forensic Sci.,* 35, 1360, 1990.
88. Pounder, D.J. and Yonemitsu, K., Postmortem absorption of drugs and ethanol from aspirated vomitus — an experimental model, *Forensic Sci. Int.,* 51, 189, 1991.
89. Corry, J.E., Possible sources of ethanol ante- and post-mortem: its relationship to the biochemistry and microbiology of decomposition, *J. Appl. Bacteriol.,* 44, 1, 1978.
90. Takayasu, T., Ohshima, T., Tanaka, N., Maeda, H., Kondo, T., Nishigami, J., and Nagano, T., Postmortem degradation of administered ethanol-d6 and production of endogenous ethanol: experimental studies using rats and rabbits, *Forensic Sci. Int.,* 76, 129, 1995.
91. Davis, G.L., Leffert, R.L., and Rantanon, N.W., Putrefactive ethanol sources in postmortem tissues of conventional and germ-free mice, *Arch. Pathol.,* 94, 71, 1972.
92. Nanikawa, R., Moriya, F., and Hashimoto, Y., Experimental studies on the mechanism of ethanol formation in corpses, *Z. Rechtsmed.,* 101, 21, 1988.
93. Bogusz, M., Guminska, M., and Markiewicz, J., Studies on the formation of endogenous ethanol in blood putrefying *in vitro*, *J. Forensic Med.,* 17, 156, 1970.
94. de Lima, I.V. and Midio, A.F., Origin of blood ethanol in decomposed bodies, *Forensic Sci. Int.,* 106, 157, 1999.
95. Moriya, F. and Ishizu, H., Can micro-organisms produce alcohol in body cavities of a living person? *J. Forensic Sci.,* 39, 883, 1994.
96. Blackmore, D.J., The bacterial production of ethyl alcohol, *J. Forensic Sci. Soc.,* 8, 73, 1968.
97. Gormsen, H., Yeasts and the production of alcohol postmortem, *J. Forensic Med.,* 1, 170, 1954.
98. Kaji, H., Asanuma, Y., Yahara, O., Shibue, H., et al., Intragastrointestinal alcohol fermentation syndrome: report of two cases and review of the literature, *J. Forensic Sci. Soc.,* 24, 461, 1984.
99. Zumwalt, R.E., Bost, R.O., and Sunshine, I., Evaluation of ethanol concentrations in decomposed bodies, *J. Forensic Sci.,* 27, 549, 1982.
100. Clark, M.A. and Jones, J.W., Studies on putrefactive ethanol production. I. Lack of spontaneous ethanol production in intact human bodies, *J. Forensic Sci.,* 27, 366, 1982.

101. Canfield, D.V., Kupiec, T., and Huffine, E., Postmortem alcohol production in fatal aircraft accidents, *J. Forensic Sci.*, 38, 914, 1993.
102. Mayes, R., Levine, B., Smith, M.L., Wagner, G.N., and Froede, R., Toxicological findings in the *USS Iowa* disaster, *J. Forensic Sci.*, 37, 1352, 1992.
103. Ball, W. and Lichtenwalner, M., Ethanol production in infected urine, *N. Engl. J. Med.*, 301, 614, 1979.
104. Gilliland, M.G.F. and Bost, R.O., Alcohol in decomposed bodies: postmortem synthesis and distribution, *J. Forensic Sci.*, 38, 1266, 1993.
105. Levine, B., Smith, M.L., Smialek, J.E., and Caplan, Y.H., Interpretation of low postmortem concentrations of ethanol, *J. Forensic Sci.*, 38, 663, 1993.
106. Nanikawa, R., Ameno, K., Hashimoto, Y., and Hamada, K., Medicolegal studies on alcohol detected in dead bodies — alcohol levels in skeletal muscle, *Forensic Sci Int.*, 20, 133, 1982.
107. Wurst, F.M., Kempter, C., Metzger, J., Seidl, S., and Alt, A., Ethyl glucuronide: a marker of recent alcohol consumption with clinical and forensic implications, *Alcohol*, 20, 111, 2000.
108. Droenner, P., Schmitt, G., Aderjan, R., and Zimmer, H., A kinetic model describing the pharmacokinetics of ethyl glucuronide in humans, *Forensic Sci. Int.*, 126, 24, 2002.
109. Helander, A. and Beck, O., Mass spectrometric identification of ethyl sulphate as an ethanol metabolite in humans, *Clin. Chem.*, 50, 936, 2004.
110. Hansson, P., Varga, A., Krantz, P., and Alling, C., Phosphatidylethanol in post-mortem blood as a marker of previous heavy drinking, *Int. J. Legal Med.*, 115, 158, 2001.
111. Refaai, M.A., Nguyen, P.N., Steffensen, T.S., Evans, R.J., et al., Liver and adipose tissue fatty acid ethyl esters obtained at autopsy are post-mortem markers for pre-mortem ethanol intake, *Clin. Chem.*, 48, 77, 2002.
112. Helander, A., Biological markers in alcoholism, *J. Neural Transm.*, 66(Suppl.), 15, 2003.
113. Helander, A., Beck, O., and Jones, A.W., Distinguishing ingested ethanol from microbial formation by analysis of urinary 5-hydroxytryptophol and 5-hydroxyindoleacetic acid, *J. Forensic Sci.*, 40, 95, 1995.
114. Helander, A., Beck, O., and Jones, A.W., Urinary 5HTOL/5HIAA as biochemical marker of postmortem ethanol synthesis, *Lancet*, 340, 1159, 1992.
115. Johnson, R.D., Lewis, R.J., Canfield, D.V., and Blank, C.L., Accurate assignment of ethanol origin in post-mortem urine: liquid chromatographic-mass spectrometric determination of serotonin metabolites, *J. Chromatogr. B*, 805, 223, 2004.
116. Lewis, R.J., Johnson, R.D., Angier, M.K., and Vu, N.T., Ethanol formation in unadulterated postmortem tissues, *Forensic Sci. Int.*, 146, 17, 2004.
117. Johnson, R.D., Lewis, R.J., Canfield, D.V., Dubowski, K.M., and Blank, C.L., Utilizing the urinary 5-HTOL/5-HIAA ratio to determine ethanol origin in civil aviation accident victims, *J. Forensic Sci.*, 50, 670, 2005.
118. Hirsch, C.S. and Adelson, L., Ethanol in sequestered hematomas, *Am. J. Clin. Pathol.*, 59, 429, 1973.
119. Buchsbaum, R.M., Adelson, L., and Sunshine, I., A comparison of post-mortem ethanol levels obtained from blood and subdura specimens, *Forensic Sci. Int.*, 41, 237, 1989.
120. Eisele, J.W., Reay, D.T., and Bonnell, H.J., Ethanol in sequestered hematomas: quantitative evaluation, *Am. J. Clin. Pathol.*, 81, 352, 1984.
121. Nanikawa, R., Ameno, K., and Hashimoto, Y., Medicolegal aspects on alcohol detected in autopsy cases — alcohol levels in hematomas [in Japanese], *Jpn. J. Leg. Med.*, 31, 241, 1977.
122. Freireich, A.W., Bidanset, J.H., and Lukash, L., Alcohol levels in intracranial blood clots, *J. Forensic Sci.*, 20, 83, 1975.
123. Smialek, J.E., Spitz, W.U., and Wolfe, J.A., Ethanol in intracerebral clot: report of two homicidal cases with prolonged survival after injury, *Am. J. Forensic Med. Pathol.*, 1, 149, 1980.
124. Cassin, B.J. and Spitz, W.U., Concentration of alcohol in delayed subdural hematoma, *J. Forensic Sci.*, 28, 1013, 1983.
125. Randall, B., Fatty liver and sudden death, *Hum. Pathol.*, 11, 147, 1980.
126. Pounder, D.J., Stevenson, R.J., and Taylor, K.K., Alcoholic ketoacidosis at autopsy, *J. Forensic Sci.*, 43, 812, 1998.
127. Kanetake, J., Kanawaku, Y., Mimasaka, S., et al., The relationship of a high level of serum betahydroxybutyrate to cause of death, *Legal Med.*, 7, 169, 2005.
128. Jenkins, D.W., Eckel, R.E., and Craig, J.W., Alcoholic ketoacidosis, *J. Am. Med. Assoc.*, 217, 177, 1971.

129. Levy, L.J., Duga, J., Girgis, M., and Gordon E.E., Ketoacidosis associated with alcoholism in nondiabetic subjects, *Ann. Intern. Med.*, 78, 213, 1973.
130. Bremer, J., Pathogenesis of ketonemia, *Scand. J. Clin. Lab. Invest.*, 23, 105, 1969.
131. Wrenn, K.D., Slovis, C.M., Minion, G.E., and Rutkowski, R., The syndrome of alcoholic ketoacidosis, *Am. J. Med.*, 91, 119, 1991.
132. Palmer, J.P., Alcoholic ketoacidosis: clinical and laboratory presentation, pathophysiology and treatment, *Clin. Endocrinol. Metab.*, 12, 381, 1983.
133. Halperin, M.L., Hammeke, M., Josse, R.G., and Jungas, R.L., Metabolic acidosis in the alcoholic: a pathophysiologic approach, *Metabolism*, 32, 308, 1983.
134. Cahill, G.F., Ketosis, *Kidney Int.*, 20, 416, 1981.
135. Isselbacher, K.H., Metabolic and hepatic effects of alcohol, *N. Engl. J. Med.*, 296, 612, 1977.
136. Pounder, D.J., Problems in the necropsy diagnosis of alcoholic liver disease, *Am. J. Forensic Med. Pathol.*, 5, 103, 1984.
137. Mihas, A.A. and Tavaossli, M., Laboratory markers of ethanol intake and abuse: a critical appraisal, *Am. J. Med. Sci.*, 303, 415, 1992.
138. Sadler, D.W., Girela, E., and Pounder, D.J., Post mortem markers of chronic alcoholism, *Forensic Sci. Int.*, 82, 153, 1996.
139. Jones, A.W. and Lowinger, H., Relationship between the concentration of ethanol and methanol in blood samples from Swedish drinking drivers, *Forensic Sci. Int.*, 37, 277, 1987.
140. Markiewicz, J., Chlobowska, Z., Sondaj, K., and Swiegoda, C., Trace quantities of methanol in blood and their diagnostic value, *Z. Zagadnien. Nauk. Sadowych.*, 33, 9, 1996.
141. Hashemy Tonkabony, S.E., Post-mortem blood concentration of methanol in 17 cases of fatal poisoning from contraband vodka, *Forensic Sci.*, 6, 1, 1975.
142. MacDougall, A.A., Clasg, M.A., and MacAulay, K., Addiction to methylated spirit, *Lancet*, Special Articles, 498, 1956.
143. Wu Chen, N.B., Donoghue, E.R., and Schaffer, M.I., Methanol intoxication: distribution in postmortem tissues and fluids including vitreous humor, *J. Forensic Sci.*, 30, 213, 1985.
144. Pla, A., Hernandez, A.F., Gil, F., Garcia-Alonso, M., and Villanueva, E., A fatal case of oral ingestion of methanol. Distribution in postmortem tissues and fluids including pericardial fluid and vitreous humor, *Forensic Sci. Int.*, 49, 193, 1991.
145. Naraqi, S., Dethlefs, R.F., Slobodniuk, R.U., and Sairere, J.S., An outbreak of acute methyl alcohol intoxication, *Aust. N.Z. Med.*, 9, 65, 1979.
146. Swartz, R.D.M., McDonald, J.R., Millman, R.P., Billi, J.E., Bondar, N.P., Migdal, S.D., Simonian, S.K., Monforte, J.R., McDonald, F.D., Harness, J.K., and Cole, K.L., Epidemic methanol poisoning: clinical and biochemical analysis of a recent episode, *Medicine*, 60, 373, 1996.
147. Shahangian, S. and Ash, K.O., Formic and lactic acidosis in a fatal case of methanol intoxication, *Clin. Chem.*, 32, 395, 1996.
148. Jacobsen, D., Jansen, H., Wiik-Larsen, E., Bredesen, J.E., and Halvorsen, S., Studies on methanol poisoning, *Acta Med. Scand.*, 212, 5, 1982.
149. Tanaka, E., Honda, K., Horiguchi, H., and Misawa, S., Postmortem determination of the biological distribution of formic acid in methanol intoxication, *J. Forensic Sci.*, 36, 936, 1991.
150. Houvda, K.E., Hunderi, O.H., Tafjord, A.B., Dunlop, O., Rudberg, N., and Jacobsen, D., Methanol outbreak in Norway 2002–2004: epidemiology, clinical features and prognostic signs, *J. Intern. Med.*, 258, 181, 2005.
151. Mani, J. C., Pietruszko, R., and Theorell, H., Methanol activity of alcohol dehydrogenase from human liver, horse liver, and yeast, *Arch. Biochem. Biophys.*, 140, 52, 1970.
152. Jacobsen, D. and McMartin, K.E., Antidotes for methanol and ethylene glycol poisoning, *Clin. Toxicol.*, 35, 127, 1997.
153. Barceloux, D.G., Bond, G.R., Krenzelok, E.P., Cooper, H., and Vale, J.A., American Academy of Clinical Toxicology practice guidelines on the treatment of methanol poisoning, *J. Toxicol. Clin. Toxicol.*, 40, 415, 2002.
154. Mycyk, M.B. and Leikin, J.B., Antidote review: fomepizole for methanol poisoning, *Am. J. Ther.*, 10, 68, 2003.
155. Brent, J., McMartin, K., Phillips, S., Aaron, C., and Kulig, K., Fomepizole for the treatment of methanol poisoning, *N. Engl. J. Med.*, 344, 424, 2001.

156. De Brabander, N., Wojciechowski, M., De Decker, K., De Weerdt, A., and Jorens, P.G., Fomepizole as a therapeutic strategy in paediatric methanol poisoning. A case report and review of the literature, *Eur. J. Pediatr.,* 164, 158, 2005.
157. Hovda, K.E., Andersson, K.S., Utdal, P., and Jacobsen, D., Methanol and formate kinetics during treatment with fomepizole, *Clin. Toxicol.,* 43, 221, 2005.
158. Adelson, L., Fatal intoxication with isopropyl alcohol (rubbing alcohol), *Am. J. Clin. Pathol.,* 38, 144, 1962.
159. Lacouture, P.G., Wason, S., Abrams, A., and Lovejoy, F.H., Acute isopropyl alcohol intoxication, *Am. J. Med.,* 75, 680, 1996.
160. Baselt, R.C. and Cravey, R.H., *Disposition of Toxic Drugs and Chemicals in Man.* 4th ed., Chemical Toxicology Institute, Foster City, CA, 1995.
161. Natowicz, M., Donahue, J., Gorman, L., Kane, M., and McKissick, J., Pharmacokinetic analysis of a case of isopropanol intoxication, *Clin. Chem.,* 31, 326, 1985.
162. Jones, A.W., Elimination half-life of acetone in humans: case-report and review of the literature, *J. Anal. Toxicol.,* 24, 8, 2000.
163. Alexander, C.B., McBay, A.J., and Hudson, R.P., Isopropanol and isopropanol deaths — ten years' experience, *J. Forensic Sci.,* 27, 541, 1982.
164. Lewis, G.D., Laufman, A.K., McAnalley, B.H., and Garriot, J.C., Metabolism of acetone to isopropyl alcohol in rats and humans, *J. Forensic Sci.,* 29, 541, 1996.

CHAPTER 4

Recent Advances in Biochemical Tests for Acute and Chronic Alcohol Consumption

Anders Helander, Ph.D.[1] **and Alan Wayne Jones, D.Sc.**[2]
[1] Department of Clinical Neuroscience, Karolinska Institute and Karolinska University Hospital, Stockholm, Sweden
[2] Department of Forensic Toxicology, University Hospital, Linköping, Sweden

CONTENTS

4.1 INTRODUCTION

Most people enjoy a drink and, for the vast majority of individuals, alcohol is a harmless, socially accepted recreational drug.[1] However, for about 10% of the population, especially among men, moderate drinking eventually leads to alcohol abuse and dependence with serious consequences for the individual and society.[2,3] Overconsumption of alcohol is a major public health hazard and a cause of premature death and morbidity.[4,5] Binge drinking, which is usually defined as consumption of five or more drinks on one occasion, is often associated with acute intoxication, hooliganism, drunk driving, and other deviant behavior with negative consequences for the person's family and friends.[6,7]

Statistics show that alcohol consumption is increasing worldwide in both sexes and this legal drug creates enormous costs for society in terms of treatment and rehabilitation of those who abuse alcohol.[8–10] Early recognition of problem drinkers in the society is therefore important to ensure adequate treatment strategies.[11] According to the American Medical Association, the differences among moderate use, abuse, and alcohol dependence ("alcoholism") are summarized as follows:

1. The consumption of alcohol in amounts considered harmless to health entails drinking at most one to two drinks per day (~10 to 20 g ethanol), and never first thing in the morning or on an empty stomach and the resulting blood alcohol concentration (BAC) should not exceed 0.2 g/L (0.02 g%) on any drinking occasion.
2. Abuse of alcohol is a pattern of drinking that is accompanied by one or more of the following problems: (a) failure to fulfill major work, school, or home responsibilities because of drinking; (b) drinking in situations that are physically dangerous, such as driving a car or operating machinery; (c) recurring alcohol-related legal problems, such as being arrested for driving under the influence of alcohol or for physically hurting someone while drunk; and (d) having social or relationship problems that are caused by or worsened by the effects of alcohol.
3. Alcohol dependence is a more severe pattern of drinking that includes the problems of alcohol abuse and persistent drinking in spite of obvious physical, mental, and social problems caused by alcohol. Also typical are (a) loss of control and inability to stop drinking once begun; (b) withdrawal symptoms associated with stopping drinking such as nausea, sweating, shakiness, and anxiety; and (c) tolerance to alcohol, needing increased amounts of alcohol in order to feel drunk.

Denial of drinking practices has always been a major stumbling block in the effective treatment of alcohol abuse and dependence.[12] Drinking histories are notoriously unreliable and this tends to complicate early detection and treatment of the underlying alcohol problem.[13,14] Much research effort has therefore focused on developing more objective ways to disclose excessive drinking, so that help can be given to those at risk of becoming dependent on alcohol.[15] In this connection, the use of various clinical laboratory tests is a useful complement to self-report questionnaires, such as the MAST and CAGE,[16,17] which are intended to divulge the quantity and frequency of alcohol consumption as well as various social-medical problems associated with alcohol abuse and dependence.

Accordingly, a multitude of biochemical markers have been developed to provide more objective ways of diagnosing overconsumption of alcohol and risk for alcohol-induced organ and tissue damage.[18–21] The liver is particularly vulnerable to heavy drinking and damage to liver cells is often reflected in an increased activity of various enzymes in the bloodstream, such as γ-glutamyl transferase (GGT) and alanine and aspartate aminotransferase (ALT and AST).[22,23] However, it seems that some individuals can drink excessively for months or years without displaying abnormal results with this kind of biochemical test, which implies a low *sensitivity* for detecting hazardous drinking. By contrast, some biological markers yield positive results in people suffering from non-alcohol-related liver problems, or after taking certain kinds of medication, which implies a low *specificity* for detecting alcohol abuse.

Nevertheless, interest in the use of biochemical tests or biomarkers for screening those individuals at most risk of developing problems with alcohol consumption has expanded greatly.[21,24,25] Besides many applications in clinical practice, such as in the rehabilitation of alcoholics and in

Table 4.1 Examples of Biochemical Markers of Alcohol Use and Abuse, and Possible Predisposition to Alcohol Dependence

Classification	Examples of Biochemical Markers
Acute Markers	Ethanol
	5-Hydroxytryptophol (5HTOL)
	Ethyl glucuronide (EtG)
	Ethyl sulfate (EtS)
	Fatty-acid ethyl esters (FAEE)
State Markers	γ-Glutamyl transferase (GGT)
	Alanine aminotransferase (ALT)
	Aspartate aminotransferase (AST)
	Mean corpuscular volume (MCV)
	Carbohydrate-deficient transferrin (CDT)
	Phosphatidylethanol (PEth)
Trait Markers	Monoamine oxidase (MAO)
	Adenylyl cyclase (AC)
	Neuropeptide Y (NPY)

drug-abuse treatment programs,[26] biochemical markers have found uses in occupational medicine,[27,28] forensic science,[29–32] and experimental alcohol research.[33,34] In general, three major classes of biochemical markers have been distinguished (examples are given in Table 4.1):

1. Tests sufficiently sensitive to detect even a single intake of alcohol, known as *acute markers* or *relapse markers*.
2. Tests that indicate disturbed metabolic processes or malfunctioning of body organs and/or tissue damage caused by long-term exposure to alcohol. This is reflected in altered hematological and/or biochemical parameters in blood or other body fluids. Such tests are referred to as *state markers* of hazardous alcohol consumption.
3. Tests that indicate whether a person carries a genetic predisposition for heavy drinking, abuse of alcohol, and development of alcohol dependence. Such tests are known as *trait markers* and often rely on identifying an abnormal enzyme or receptor pattern at the molecular level. Those prone to develop into heavy drinkers exhibit at an early age marked personality disorders, including sensation-seeking behavior, binge drinking, and abuse of other drugs.

In this chapter, we present an update of research dealing with laboratory markers for both acute and chronic drinking. The advantages and limitations of various laboratory tests are discussed and suggestions are made for their rational use in clinical and forensic medicine.

4.2 DIAGNOSTIC SENSITIVITY AND SPECIFICITY

Biochemical markers are usually evaluated in terms of diagnostic sensitivity and specificity. *Sensitivity* refers to the ability of a test to detect the presence of the trait in question, whereas *specificity* refers to its ability to exclude individuals without the trait. Consequently, a marker with high sensitivity yields relatively few false-negative results and one with high specificity gives few false positives. The ideal marker should, of course, be both 100% sensitive and specific, but this is never achieved because reference ranges for normal and abnormal values always tend to overlap. Instead, a cutoff, or threshold limit, is established for what is considered normal. These limits are usually determined empirically as the mean plus or minus two standard deviations (SD) of the test results for a healthy control population. Accordingly, 2.5% of individuals will be above the upper limit and 2.5% below the lower limit and the test specificity will always be less than 100%.

To obtain a sufficiently high specificity for routine purposes, the sensitivity of some markers has to be gradually reduced. On the other hand, most tests aimed at indicating liver damage caused by prolonged alcohol abuse often suffer from low specificity, because many liver diseases have non-alcoholic origin. So-called receiver-operating characteristic (ROC) curves are widely used for evaluating utility of biochemical markers and for comparing different analytical methods.[35] ROC curves are graphic illustrations created by plotting the relation between sensitivity (i.e., the percentage of true positives) against 1-specificity (i.e., the percentage of false positives) at different cutoff limits between normal and abnormal values.[36]

Most studies aimed at evaluating the sensitivity and specificity of alcohol biomarkers rely heavily on patient self-reports about drinking as the gold standard. However, considering that many patients fail to provide an accurate history of their true alcohol consumption, this creates a validity problem. Hence, besides the use of sensitive and specific markers of excessive alcohol consumption, there is also a need to develop and evaluate laboratory tests to monitor recent alcohol consumption in a more objective way.

4.3 TESTS FOR ACUTE ALCOHOL INGESTION

4.3.1 Measuring Ethanol in Body Fluids and Breath

Ethanol and water mix together in all proportions and, after drinking alcoholic beverages, the ethanol distributes into all body fluids and tissues in proportion to the amount of water in these fluids and tissues. The body water in men makes up about 60% of their body weight and the corresponding figure for women is ~50%, although there are large inter-individual differences in these average figures, depending on age and, especially, the amount of adipose tissue. Accordingly, the most specific and direct way to demonstrate that a person has been drinking alcohol is to analyze a sample of blood, breath, urine, or saliva. However, because concentrations of ethanol in these body fluids decrease over time, owing to metabolism and excretion processes, the time frame for positive identification is rather limited.[37,38]

The smell of alcohol on the breath is perhaps the oldest and most obvious indication that a person has been drinking. But many alcoholics use breath fresheners or can regulate their intake so that the BAC is low or zero when they are examined by a physician.[39] A more objective way to disclose recent alcohol consumption is to measure the concentration of ethanol in the exhaled air. Several kinds of handheld breath alcohol analyzers are available for this purpose, such as Alcolmeter SD-400, AlcoSensor IV, or Alcotest 4010. The ethanol in a sample of breath is oxidized with an electrochemical sensor and the magnitude of the response is directly proportional to the concentration of ethanol present.[40] Studies have shown that these breath analyzers are accurate, precise, and selective for their intended purpose. Endogenous breath volatiles, such as acetone, are not oxidized under the same conditions and therefore do not interfere with the selectivity of the test for ethanol.

Breath alcohol concentration (BrAC) tests should become a standard procedure if a patient is required to refrain from drinking as part of rehabilitation or treatment or because of workplace regulations concerning the use of alcohol.[41] However, a positive breath test needs to be confirmed by making a repeat test not less than 15 min later, to rule out the presence of ethanol in the mouth from recent drinking. Most of the currently available handheld breath alcohol analyzers have an analytical sensitivity of about 0.05 mg ethanol per liter breath, which corresponds to a blood ethanol equivalent of 10 mg/dL (~2.2 mmol/L). The result of a breath alcohol test appears immediately after capturing the sample and results are reported in units of g/210 L (U.S.) or mg/L (Sweden) or μg/100 mL (U.K.). Alternatively, the result of the test is translated into the presumed coexisting BAC and for this application the breath alcohol instrument is precalibrated with a blood/breath conversion factor, usually assumed to be 2100:1 or 2300:1. Careful control of calibration and maintenance of these breath test instruments is important to ensure obtaining valid and reliable results.

Measuring the concentration of ethanol in whole blood or plasma/serum will also provide reliable information about recent drinking. However, obtaining a sample of blood is an invasive procedure and the concentration of ethanol, if any, is not obtained immediately after sampling. The analysis of ethanol in blood or plasma is therefore less practical than breath testing, for clinical purposes, as a rapid screening test for recent drinking. The sensitivity of methods for blood alcohol analysis (e.g., gas chromatography; GC) is higher than breath test instruments and a BAC as low as 1 mg/dL can be measured. However, for clinical applications, it is wise to use a higher cutoff (i.e., decision limit) such as 5 or 10 mg/dL, to avoid discussions and debate that the ethanol came from some dietary constituent, such as fresh fruits or soft drinks.

After absorption and distribution of ethanol in body fluids and tissues is complete, there is a close correlation between the concentrations in saliva, blood, and urine. The equilibration of ethanol between blood and saliva is fairly rapid, which makes saliva sampling more suitable than urine for clinical purposes.[42,43] A number of devices have been developed for measuring ethanol in saliva and these have proved useful for alcohol screening purposes in clinical settings. A saliva-test device called QED has been evaluated extensively and gives on-the-spot results as to whether a person has consumed alcohol. The QED test incorporates alcohol dehydrogenase (ADH) to oxidize ethanol with the coenzyme NAD^+ at pH 8.6. Ethanol is converted into acetaldehyde and the NADH is formed in direct proportion to the concentration of ethanol present. The acetaldehyde is trapped with semicarbazide to drive the reaction to completion. The NADH is then re-oxidized to produce a colored end product, by reaction with the enzyme diaphorase and a tetrazolium salt incorporated on a solid phase support. The length of the resulting blue-colored bar is directly proportional to the concentration of ethanol in the saliva sample and permits a direct readout of the test result about 1 min later. Saliva alcohol concentrations determined with QED agreed well with BAC and BrAC in controlled drinking experiments.[44,45]

Numerous studies have compared concentrations of ethanol in blood and urine sampled at various times after end of drinking.[46,47] In the post-absorptive phase, the urine-alcohol concentration (UAC) and the BAC are highly correlated ($r > 0.95$). Some have tried to estimate BAC indirectly from UAC, assuming a population average UAC/BAC, such as 1.3:1. However, there are large inter- and intra-individual variations in this relationship, which means that the estimated BAC will have a considerable uncertainty in any individual case.

One expects to find a higher concentration of ethanol in urine compared with blood because of the difference in water content of these body fluids, namely, 100% vs. 80%. This suggests a UAC/BAC ratio of 1.25:1 for freshly produced urine. In reality, however, the UAC/BAC ratio also depends on the time after drinking, when the bladder was last voided, and how frequently the person urinates. Urine is stored but not metabolized in the bladder, whereas the BAC changes continuously, depending on the stage of metabolism and the rate of hepatic oxidation. Shortly after drinking during the absorption phase, the UAC and BAC are not well correlated, whereas in the post-peak phase, when BAC is decreasing at a constant rate of about 15 mg/dL/h, a good correlation exists between BAC and UAC.

The average curves for venous blood and urine concentration–time profiles of ethanol are compared in Figure 4.1. One notes that UAC and BAC curves are shifted in time, as a consequence of the time-lag between ethanol being absorbed into the bloodstream, reaching the kidney, and passing into the glomerular filtrate, and its storage in the bladder until voided. Shortly after the end of drinking, the UAC is less than the BAC (UAC/BAC < 1.0). After the peak BAC is reached, the two curves cross and the UAC has a higher C_{max} compared with the BAC. In the post-absorptive phase, the UAC is always higher than the corresponding BAC by a factor of 1.3 to 1.4. Note that the UAC reflects the average BAC prevailing during the time that urine was produced and stored in the bladder since the previous void. The UAC in a random void does not reflect the BAC at the time of emptying the bladder and, in this respect, is less useful than blood, saliva, or breath as a test of alcohol influence. Instead, the UAC reflects the BAC during production and storage of urine in the bladder. The UAC remains elevated for about 1 h after the BAC has already reached zero. Accordingly, the first morning void after an evening's drinking might be positive for ethanol, although the concentrations in blood or breath have already reached zero.[48] This relationship

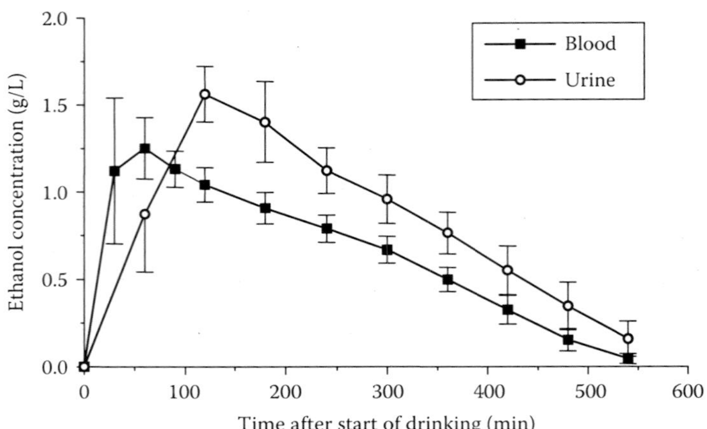

Figure 4.1 Mean concentration-time profiles of ethanol in blood and urine in 30 healthy men who drank 0.85 g/kg body weight after an overnight fast. The bladder was emptied before the start of drinking and alcohol was taken in the form of neat whisky.

suggests that the BAC has reached zero sometime during the night and any ethanol already in the urine gets diluted with ethanol-free urine produced after complete metabolism of the alcohol consumed. Metabolism of ethanol does not occur in the urinary bladder, and back-diffusion of ethanol into the bloodstream is negligible, owing to the limited blood circulation.

Small quantities of ethanol are excreted through the skin by passive diffusion and also secreted through the sweat glands. The transdermal elimination of ethanol corresponds to about 0.5 to 1% of the dose ingested.[49] However, this route of excretion has found applications in clinical medicine as a way to monitor alcohol consumption over periods of several weeks or months. This approach might be useful to control if alcoholics and others manage to remain abstinent, and has led to the introduction of a procedure known as transdermal dosimeter or, more simply, the sweat-patch test.[50,51] Although the first attempts to monitor alcohol consumption in this way were not very successful, owing to technical difficulties with the equipment used for collecting sweat, the procedures are now much improved and can be used to analyze other drugs of abuse as well.[52,53] The test person wears a tamper-proof and water-proof pad, positioned on an arm or leg, and the low-molecular substances that pass through the skin are collected during the time the patch remains intact. Ethanol and other volatiles are extracted with water and the concentration determined provides a cumulative index of alcohol exposure. The ethanol collected in the cotton pad can be determined in a number of ways, such as by extraction with water and GC analysis or by headspace vapor analysis with a handheld electrochemical sensor, which was originally designed for breath alcohol testing.[54] A miniaturized electronic device for continuous sampling and monitoring of transcutaneous ethanol has recently been introduced.[55,56]

4.3.2 Metabolism of Ethanol

The disposition and fate of ethanol in the body have been studied extensively since the 1930s and our knowledge about this legal drug exceeds that of other abused substances. Ethanol is cleared from the bloodstream by both oxidative and non-oxidative metabolic pathways (Figure 4.2).

The minor non-oxidative pathway of alcohol metabolism has received considerable research interest since the first edition of *Drug Abuse Handbook* appeared and this topic is covered later in this chapter. The main alcohol-metabolizing enzymes are located in the liver, the kidney, and the gastric mucosa. The bulk of the alcohol a person consumes undergoes hepatic metabolism by the action of Class I ADH, which exists in various molecular forms, so-called isozymes. Ethanol is

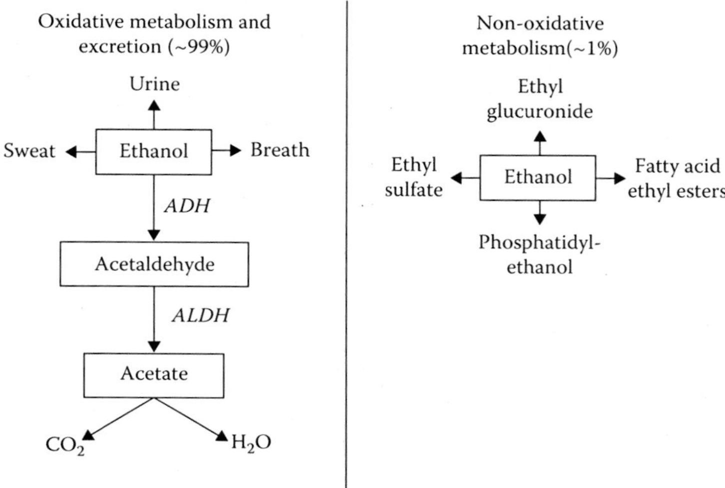

Figure 4.2 Fate of alcohol in the body illustrating both the oxidative and non-oxidative pathways of ethanol metabolism.

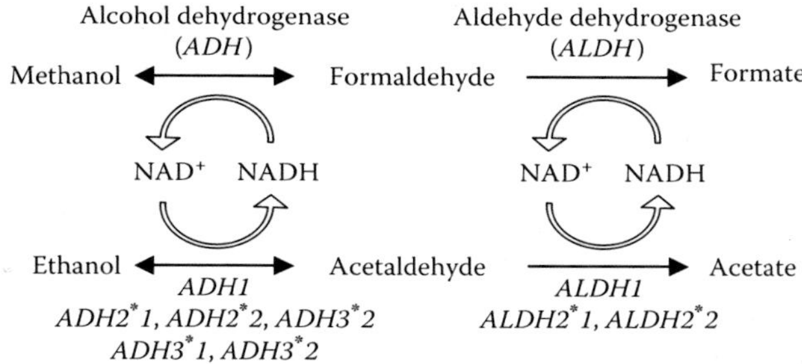

Figure 4.3 Schematic diagram comparing the metabolism of ethanol and methanol via the alcohol dehydrogenase pathway: NAD^+ = oxidized form of the coenzyme nicotinamide adenine dinucleotide; NADH = reduced form of the coenzyme. The various isozymes of alcohol dehydrogenase (ADH) and aldehyde dehydrogenase (ALDH) are indicated.

metabolized in a two-stage process, first to acetaldehyde, and this primary metabolite is rapidly converted to acetate (acetic acid) by the action of low K_m aldehyde dehydrogenase (ALDH2) located in the mitochondria. The end products of the oxidation of ethanol are carbon dioxide and water (see Figure 4.2).

Hepatic ADH is not specific for oxidation of ethanol, and other aliphatic alcohols, if present in the blood, as well as a number of endogenous substances (e.g., prostaglandins and hydroxysteroids), also serve as substrates. The substrate specificity of ADH toward aliphatic alcohols differs widely, and the rate of oxidation of methanol is considerably slower than that of ethanol by a factor of about 10:1.[57] The biotransformation of ethanol and methanol and the various metabolic products formed are compared in Figure 4.3.

Raised concentrations of the intermediary products of ethanol oxidation have been proposed as a way to test for recent drinking.[56] However, measuring acetaldehyde is not very practical because of the extremely low concentrations present (<1% of the ethanol concentration), and also the fact that the necessary analytical procedures are much more challenging than those for the analysis of ethanol.[58] Acetaldehyde is rapidly converted to acetate, and the concentration of free acetaldehyde in peripheral venous plasma is further reduced owing to a more or less specific binding to various endogenous molecules such as proteins (see also Section 4.4.6). An additional problem arises if the blood contains ethanol, because acetaldehyde is formed after sampling resulting in falsely high results.[58,59] Measuring acetaldehyde in breath instead of blood has been suggested as an alternative approach, although even breath testing is not without its problems.[60]

The concentration of acetate in blood depends on the rate of ethanol oxidation and utilization of the acetate formed by peripheral tissues. The blood acetate concentration appears to be independent of the blood-ethanol concentration, and instead increases with the development of metabolic tolerance to alcohol (i.e., rate of ethanol elimination).[61,62] Measuring blood acetate was suggested as a marker of chronic abuse of alcohol,[63,64] and the sensitivity and specificity of this test was significantly higher than for GGT.[63] It should be emphasized, however, that blood-acetate remains elevated only as long as ethanol is being metabolized and, moreover, the rate of ethanol metabolism exhibits large inter-individual variations even in moderate drinkers.

4.3.3 Analysis of Methanol in Body Fluids

Ethanol and methanol are examples of endogenous alcohols and are normally present in biological specimens, albeit at extremely low concentrations (<1 mg/L). Moreover, trace quantities of the free alcohols or their esters might be ingested with certain foodstuffs or might be contained in soft drinks, such as fruit juices and sodas, or they could be formed by fermentation of dietary carbohydrates through the action of microorganisms inhabiting the gut.[65–68]

During the end stages of carbohydrate metabolism, trace amounts of acetaldehyde are produced from pyruvate and, after reduction via the ADH/NAD$^+$ pathway, lead to small amounts of ethanol being present in body fluids.[69] The trace amounts of the endogenous alcohols produced in the gut are rapidly cleared from the portal venous blood as it passes through the liver for the first time. The existence of an effective first-pass metabolism ensures that only vanishingly small concentrations of ethanol and methanol reach the peripheral circulation. With the use of highly sensitive and specific analytical methods, the concentrations of ethanol and methanol in body fluids obtained from healthy abstaining individuals range from 0.04 to 0.1 mg/dL.[70,71]

Ethanol and methanol compete for binding sites on the Class I isozymes of ADH, which show a stronger preference for the oxidation of ethanol.[57] As a consequence, during metabolism of ethanol, the concentration of methanol in blood increases and remains on a more or less constant level, until blood ethanol decreases below 20 mg/dL (~4.3 mmol/L) (Figure 4.4). Thereafter, methanol is cleared with a half-life of 2 to 3 h, which means that methanol can be detected in body fluids long after the concentration of ethanol has returned to baseline or endogenous levels.[37,72–74] This protracted wash-out of methanol opens the possibility of verifying recent drinking for several hours after ethanol has been cleared from the body.

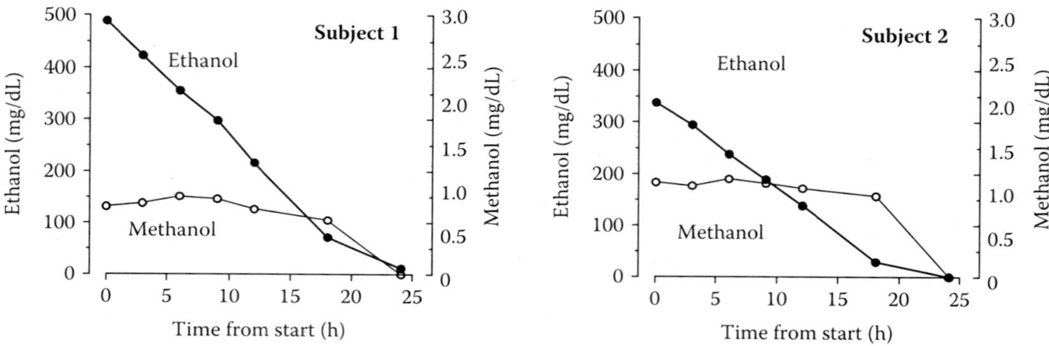

Figure 4.4 Elimination kinetics of ethanol and methanol in two alcoholics during detoxification. Note that the concentration of methanol in blood remains fairly constant at about 1 mg/dL until the ethanol concentration decreases toward zero.

Analysis of blood and urine methanol is included in some forensic investigations, when accountability for road traffic and workplace accidents are investigated.[38] Besides the acute effects of alcohol on performance and behavior, many people are impaired the morning after an evening's heavy drinking; these post-intoxication effects of heavy drinking are known as hangover (see Chapter 1).

The higher affinity of ADH for oxidation of ethanol compared with methanol also explains the mechanism behind the therapeutic usefulness of ethanol in treating methanol poisoning. Ethanol is given by intravenous infusion to reach and maintain a blood ethanol concentration of about 100 to 120 mg/dL, which effectively blocks the metabolism of methanol into its toxic metabolites formaldehyde and formic acid.[75] In the meantime, the more dangerous methanol is removed from the blood by dialysis and bicarbonate can be administered to counteract metabolic acidosis.[76] However, there are some reservations against the continued use of this antidote, because ethanol exerts a pronounced effect on the central nervous system and should not be used in treating children who inadvertently ingest methanol (wood alcohol).[77] The antidote currently in vogue for treating people poisoned with methanol is 4-methyl pyrazole (fomepizole), a competitive inhibitor of ADH.[78,79]

Although alcoholic beverages are primarily mixtures of ethanol and water, they also contain a multitude of other chemical compounds, albeit at extremely low concentrations. These other substances, which are produced as by-products of the fermentation process, are collectively known as congeners and impart the smell and flavor to the alcoholic beverage.[80] Methanol is a ubiquitous congener and is present in beer (0.1 to 1 mg/dL), red and white wines (2 to 10 mg/dL), gin and whisky (0.1 to 20 mg/dL), and brandies and cognac (20 to 200 mg/dL).[81] Accordingly, a raised concentration of methanol in blood and other body fluids after drinking alcoholic beverages could partly be explained by its presence as a congener, and also because of the metabolic interaction with ethanol via competitive inhibition of ADH.[82] Finding a blood methanol concentration above 1 mg/dL suggests accumulation during a period of long-term heavy drinking (days or weeks), and some feel this furnishes a test for chronic alcohol abuse (see also Section 4.4.7).[83–85] However, care is needed to ensure that the alcoholic beverage consumed does not contain abnormally high concentrations of methanol.[81,86,87]

4.3.4 Conjugates of Ethanol Metabolism

The glucuronidation pathway of drug metabolism, including the conjugation of hydroxyl groups in aliphatic alcohols, was discovered more than a century ago.[88] Shortly after, it was found that ethanol was converted into ethyl glucuronide (EtG) and this minor metabolite was identified in urine.[89] However, interest in this non-oxidative pathway of ethanol metabolism was hampered, owing to the complexity of the analytical methods needed to determine reliably EtG concentrations in body fluids.[90,91] The advent of modern analytical techniques, such as the highly sensitive and specific gas and liquid chromatography combined with mass spectrometry (GC-MS and LC-MS),[92] radically changed the situation and spawned in a new area of biomedical alcohol research. Careful studies showed that only a very small fraction of the ingested ethanol (<0.1%) undergoes phase II conjugation reactions with glucuronic acid or sulfuric acid to produce EtG and ethyl sulfate (EtS), respectively (Figure 4.5).[93,94] These soluble non-oxidative metabolites of ethanol are excreted with the urine, for several hours longer than ethanol itself.[93,95–97]

After acute alcohol ingestion, there is no accumulation of EtG in body fluids,[98] although it may be retained in hair.[99] The advantage of hair over other body media for drug analysis is the extended window of detection offered, and, based on segmental analysis and an average growth rate, some idea can be obtained as to when the drug was last being used.[100] Determination of EtG in urine provides a means to verify whether a person has recently been drinking alcohol, for much longer than ethanol is detectable in body fluids. The resulting higher sensitivity of the EtG test,

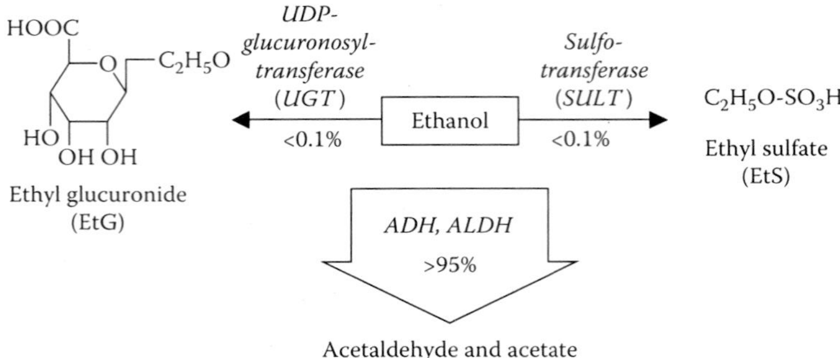

Figure 4.5 Schematic diagram illustrating the relative importance of the oxidation (phase I) and non-oxidative conjugation (phase II) elimination pathways for ethanol in the human body.

compared with analysis of ethanol itself, has obvious practical advantages as an acute marker of alcohol consumption.

The availability of improved analytical methods for EtG based on mass spectrometry (GC-MS and LC-MS)[95,101,102] has allowed several research groups to make detailed studies of this trace metabolite of ethanol as a sensitive and specific biomarker of acute alcohol consumption.[21,103] However, as expected from knowledge of glucuronides in general, EtG is sensitive to urine dilution. The EtG concentration in urine can be lowered markedly by drinking large amounts of fluid prior to voiding, whereas this does not influence the EtG/creatinine ratio.[93] Furthermore, EtG is sensitive to enzymatic hydrolysis, so if biological samples are contaminated with certain bacteria containing β-glucuronidase (e.g., *Escherichia coli* which is common in urinary tract infections), this represents another reason for obtaining falsely low or false-negative results.[104]

Another phase II metabolic pathway, namely, sulfate conjugation, produces EtS and this minor metabolite of ethanol (<0.1%) is also excreted in urine.[94,97] EtS shows similar excretion kinetics to EtG,[94,97,105,106] but, in contrast to EtG, it is apparently not sensitive to bacterial hydrolysis.[104] Other minor metabolites of ethanol with potential use as alcohol biomarkers include ethyl nitrite and ethyl phosphate.[107]

4.3.5 Fatty-Acid Ethyl Esters

Fatty acid ethyl esters (FAEE), such as ethyl palmitate and ethyl oleate, are esterification products of fatty acids and ethanol synthesized through the action of the enzyme FAEE synthase. The presence of these short-chain esters in samples of blood, tissue, or hair has been proposed as a biomarker of alcohol intake with clinical and forensic applications.[100,108–110] After alcohol intake, the serum concentration of FAEE initially closely parallels that of ethanol (e.g., similar time for peak concentrations) but, because of a very slow terminal elimination phase, FAEE persists in the blood for some time after ethanol is no longer detectable.[111,112] The elimination rate of FAEE is faster than for some other acute alcohol biomarkers, implying a lower sensitivity for recent drinking.[112,113] Increasing evidence indicates that FAEE may be toxic metabolites and, as such, possible mediators of ethanol-induced organ damage.[114]

4.3.6 Metabolites of Serotonin

Studies have shown that the metabolic interaction between ethanol and serotonin (5-hydroxytryptamine) can help to detect recent alcohol consumption. 5-Hydroxytryptophol (5HTOL) is normally a minor metabolite of serotonin, but the proportion of this metabolite increases dramatically in a dose-dependent manner after drinking alcohol. At the same time, 5-hydroxyindole-3-

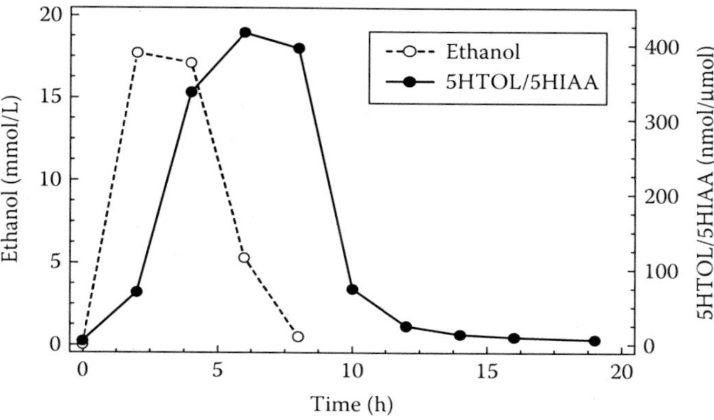

Figure 4.6 Time course of concentrations of ethanol and the ratio of 5-hydroxytryptophol (5HTOL) to 5-hydroxyindole-3-acetic acid (5HIAA) in urine after a healthy volunteer drank a single moderate dose of ethanol. The longer recovery time for the 5HTOL/5HIAA ratio is clearly evident.

acetic acid (5HIAA), the major metabolite under normal conditions, is concomitantly decreased.[115,116] Experiments with liver homogenates suggest that the shift in serotonin metabolism occurs because of competitive inhibition of ALDH by the acetaldehyde derived from oxidation of ethanol.[117] Furthermore, during the metabolism of ethanol and acetaldehyde, the reduced coenzyme NADH is present in excess and the redox state of the liver shifts to a more reduced potential, both in the cytosol and mitochondria compartments. This alters the equilibrium between several other endogenous NAD-dependent reactions, such as lactate/pyruvate and β-hydroxybutyrate/acetoacetate. The combined influences of a competitive inhibition of ALDH and altered redox state of the liver promote formation of 5HTOL at the expense of 5HIAA.[118,119] Most importantly, however, the urinary excretion of 5HTOL will not normalize until several hours after blood and urinary ethanol reach zero (Figure 4.6).[116] On the basis of this time lag, an increased urinary concentration of 5HTOL was suggested and used as a sensitive biochemical marker of recent drinking.[120–122]

Expressing 5HTOL as a ratio 5HTOL/5HIAA, rather than 5HTOL/creatinine, improves test accuracy because dietary serotonin (high amounts in banana, pineapple, kiwi fruit, and walnuts) might otherwise cause false-positive results, owing to a general increase in the urinary output of both serotonin metabolites.[123] This device also compensates for variations in the concentration of 5HTOL caused by urine dilution after drinking fluids. To discriminate between a normal and elevated urinary ratios of 5HTOL/5HIAA, a cutoff limit of 15 nmol/µmol (i.e., 1.5%) is recommended for use in clinical practice.[122] This threshold value is based on studies with alcohol-free subjects of both Caucasian and Asian origin.[124] The 5HTOL/5HIAA ratio remains stable both within-days and between-days during periods of abstinence, and the metabolites are also relatively uninfluenced during transport, handling, and long-term storage of the urine specimens. Neither gender nor genetic variations in ADH and ALDH isozyme patterns seem to influence the baseline ratio of 5HTOL/5HIAA.[37,124]

An increased urinary 5HTOL/5HIAA furnishes a specific and more sensitive laboratory test of recent drinking than measuring the concentration of ethanol or methanol in body fluids.[37,74] The major advantage of 5HTOL/5HIAA over ethanol and methanol is that a raised serotonin metabolite ratio persists for several hours longer, thereby improving the ability to detect covert drinking. Furthermore, in contrast to methanol, the baseline ratio of 5HTOL/5HIAA is not elevated after prolonged intermittent alcohol intake and can therefore identify recent drinking in moderate as well as chronic drinkers.[98] Apart from alcohol ingestion, the only factor known to increase the 5HTOL/5HIAA ratio is treatment with the ALDH inhibitors disulfiram (Antabuse) and cyanamide.[125] Another drawback is that urinary 5HTOL testing at present requires rather sophisticated analytical techniques based on GC-MS or LC-MS.[122,126,127]

Testing of urinary 5HTOL has, for example, been used in clinical practice, to detect alcohol use and single lapses during treatment of alcohol-dependent subjects in an outpatient setting,[128,129] in heroin addicts on methadone maintenance,[39] and in surgical patients with chronic alcohol abuse.[130] Furthermore, testing for 5HTOL/5HIAA has applications in forensic medicine, to distinguish ingested from microbially formed ethanol, which might occur in post-mortem specimens or in urine from individuals with diabetes or others with urinary tract infections.[29,30,32,131] The 5HTOL/5HIAA ratio has also been recommended for use during official forensic investigations of aviation crashes,[132] where the risk for post-mortem synthesis of ethanol is exaggerated.[126,133]

4.4 TESTS OF CHRONIC ALCOHOL INGESTION

4.4.1 Gamma-Glutamyl Transferase

For many years, the activity of the enzyme gamma-glutamyl transferase (GGT) in serum has been the most widely used biochemical test for alcohol abuse.[134] GGT is a membrane-bound glycoprotein widely distributed in various organs, and plays an important role in glutathione synthesis, amino acid transport, and peptide nitrogen storage. Only trace amounts are normally present in serum from healthy subjects. However, in liver damage, for example, resulting from continuous heavy drinking, a significant elevation in serum GGT occurs.[135,136] Although the mechanisms responsible are not known exactly, damage to hepatocytes and/or induction of hepatic GGT may cause the enzyme to leak into the blood.[137,138] After withdrawal from ethanol, GGT returns to normal levels within approximately 4 to 5 weeks.

The determination of GGT is routinely included in blood-chemistry profiles on admission to hospital. The major disadvantage of GGT is that the serum level is raised by a variety of other conditions besides alcohol misuse, thereby reducing its diagnostic specificity.[134,139,140] For example, several common medications, such as barbiturates and antiepileptics, and various liver disorders of non-alcohol origin, also elevate serum GGT. Moreover, accepted normal ranges depend on nutritional status, body weight, age, and gender of the individual. Different threshold limits depicting abnormal values need to be applied for women and men separately. Although GGT has limited utility as a single screening test for hazardous alcohol consumption in nonselected populations,[141] many of the confounding factors are well known and can often be excluded or controlled for in clinical situations. The major advantage of GGT as a marker of alcohol abuse is its ready availability at low cost from most clinical laboratories.

4.4.2 Aspartate and Alanine Aminotransferase

Other standard tests of liver dysfunction caused by hazardous drinking include raised serum levels of the transaminases aspartate (AST) and alanine aminotransferase (ALT), which are enzymes involved in amino acid metabolism. Like GGT, certain medical conditions other than alcohol abuse can cause abnormal AST and ALT values, and both these markers are typically less sensitive, though somewhat more specific, than GGT.[142,143] The ALT-to-AST ratio,[139] as well as the proportion of mitochondrial to total AST,[144,145] has been suggested as a way to discriminate alcohol-induced from non-alcoholic liver disease, but this has not been confirmed in studies in unselected populations.[146,147] The transaminases may be useful in combination with other biochemical tests,[148] and for follow-up of patients with already established alcoholic liver disease.[149]

4.4.3 Erythrocyte Mean Corpuscular Volume

Mean corpuscular volume (MCV) is included as part of a routine blood count and indicates the size of the red blood cells (erythrocytes). An elevated MCV is often observed in alcoholic

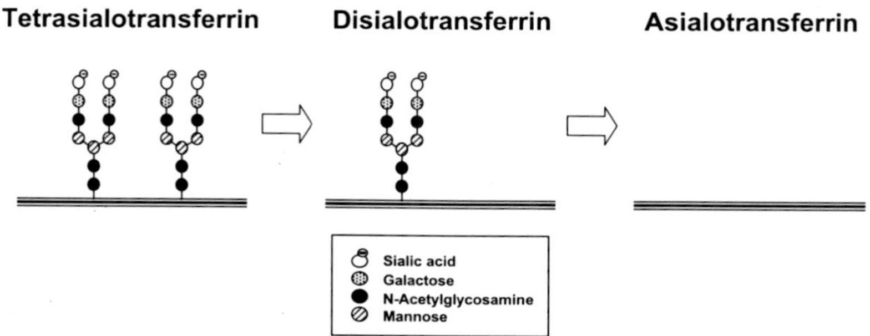

Figure 4.7 Schematic illustration of the changes in transferrin glycosylation pattern caused by heavy alcohol consumption. Tetrasialotransferrin is normally the major glycoform (~75 to 80%), but prolonged heavy drinking results in increased levels of transferrin molecules lacking one or both of the entire *N*-glycans (disialo- and asialotransferrin, respectively).

patients,[150] and this parameter has been widely used as a marker of excessive alcohol consumption.[151] The underlying cause of swelling of the red cells is unknown, but may be a direct toxic effect of ethanol or the alcohol-mediated deficiency of folic acid. The sensitivity of MCV as a biochemical marker of heavy drinking is much too low to motivate its use as a sole indicator of this condition.[152] There are other explanations for elevated MCV besides continuous heavy drinking, such as smoking, which is very common in alcoholics.[153,154] However, MCV shows a higher specificity than GGT,[148,151,155,156] and it is often used in combination with other biochemical parameters and offers the added advantage that it takes longer (several months) for MCV to recover to normal values after cessation of heavy drinking.

4.4.4 Carbohydrate-Deficient Transferrin

The presence of an abnormal glycoform pattern of the iron transport glycoprotein transferrin in serum, named carbohydrate-deficient transferrin (CDT), has emerged as the most reliable routine method for detection of continuous high alcohol consumption for 2 or more weeks.[157] The discovery of CDT as a marker of heavy drinking dates back more than 20 years.[158] The finding of an abnormal transferrin pattern in blood serum appears to be fairly specific for overconsumption of alcohol and recovers to normal during periods of abstinence with a half-life of ~1.5 to 2 weeks.[158,159]

Normal variations in transferrin glycoform patterns could stem from genetic polymorphism, the degree of iron saturation, or the number of terminal sialic acid residues in the two *N*-linked oligosaccharide chains (*N*-glycans).[160–162] The most abundant glycoform in serum is tetrasialotransferrin, containing two biantennary carbohydrate chains making a total of four terminal sialic acid residues. Studies have shown that after prolonged heavy drinking, the abundance of transferrin molecules lacking one or both of the entire *N*-glycans (disialo- and asialotransferrin, respectively) increases appreciably (Figure 4.7).[163–165] The underlying mechanism that causes elevation of CDT may involve acetaldehyde-mediated inhibition of the enzymes responsible for glycosyl transfer.[166–168]

CDT is a much more specific indicator of excessive drinking than any of the other currently used laboratory tests, and it is also assumed to have better sensitivity for early detection of alcohol abuse compared to GGT.[158,169,170] However, it should be noted that a person typically needs to drink ~50 to 80 g ethanol or more daily over a period of at least 1 to 2 weeks before abnormal CDT values develop.[158,171,172] Falsely high or low CDT results are obtained in cases of uncommon genetic transferrin variants,[162,173–175] and in rare congenital disorders of glycosylation (CDG).[176,177] It is generally agreed that the CDT content should be expressed in relation to total transferrin (i.e., %CDT), instead of as an absolute amount (e.g., mg/L). This device has an analytical advantage, because an abnormally high or low serum transferrin concentration may otherwise render falsely high or low CDT results.[178–180]

Because the disialo- and asialotransferrin glycoforms lacking 2 or 4 terminal sialic acid residues are less negatively charged, and thereby have higher isoelectric points than tetrasialotransferrin (pI ~ 5.4), they are readily separated by techniques such as isoelectric focusing (IEF), capillary electrophoresis (CE), and ion-exchange chromatography (e.g., HPLC).[159,161,181–183] Although all these methods are laborious and rather time-consuming for routine use, especially when a large number of samples must be analyzed daily, an advantage is the visible documentation and, therefore, the reduced risk of obtaining false-positive results owing to genetic variations or other chromatographic interferences.[162] The first direct immunoassay for determination of %CDT was recently introduced, which at the same time appears to be insensitive to genetic transferrin variants.[184,185]

4.4.5 Phosphatidylethanol

Another non-oxidative pathway of ethanol elimination is the reaction with membrane phospholipids to produce phosphatidylethanol (PEth). PEths are unique phospholipids formed in cell membranes only in the presence of ethanol,[186,187] and have been proposed as a highly specific marker of chronic heavy drinking within the previous week(s). After alcoholic patients, with a self-reported daily intake of 60 to 300 g ethanol for at least 1 week, were admitted to a detoxification program, PEth was detectable in their blood for another 2 weeks.[188,189] Moreover, the sensitivity of PEth was reported to be greater than, or at least equal to, that of CDT.[190] Improved analytical procedures for quantitation of PEth will allow for routine clinical use of this biomarker.[191] However, a recently recognized problem, interfering with the specificity of PEth as a biomarker of prolonged heavy drinking, is the risk of artifact formation of PEth during storage if the blood sample contains ethanol.[192]

4.4.6 Acetaldehyde Adducts

Acetaldehyde, the proximate metabolite of ethanol oxidation, is a highly reactive chemical species and forms adducts with various biomolecules, including DNA, phospholipids, and proteins.[193–195] The binding of acetaldehyde to hepatocellular macromolecules has been suggested as the underlying cause of alcoholic liver injury.[196–198] Measurements of "whole blood-associated" acetaldehyde, acetaldehyde-hemoglobin adducts, or antibodies that recognize acetaldehyde-modified structures have emerged as possible biochemical markers of excessive drinking.[199–202] However, some of these positive results were obtained *in vitro* after nonphysiological concentrations of acetaldehyde were administered, and the relevance of these findings *in vivo* has been a matter of debate. Although analysis of various forms of bound acetaldehyde have shown promising results, and, indeed, were found to be more sensitive for the detection of chronic excessive drinking than GGT and MCV,[203,204] more work is needed to determine the reliability and diagnostic potential for identification of heavy drinking in unselected populations. Moreover, most of the currently used methods for quantification of acetaldehyde adducts are probably too complex for routine purposes.

4.4.7 Other Potential Tests or Markers of Chronic Drinking

Numerous other candidate biochemical tests of alcohol use and abuse have been evaluated over the years, but only a few have gained general acceptance.[18,19] The major deficiencies are either poor sensitivity or specificity for alcohol abuse, or requiring laborious and time-consuming assay procedures. Some markers have not been tested in sufficiently large enough cohorts of subjects with use of appropriate control groups.

Several studies have reported that alcoholics after a recent or ongoing chronic drinking binge develop a lower blood, or erythrocyte, ALDH activity compared to healthy controls.[205] However, there is a considerable inter-individual overlap between heavy and moderate consumers of alcohol in this respect, and even between heavy drinkers and teetotalers.[206,207] Consequently, to obtain sufficient sensitivity, the specificity of this enzyme marker must be very low (see Section 4.2). In addition, several drugs, including the alcohol-sensitizing agents disulfiram (Antabuse) and cyanamide, as well as environmental factors like smoking, may cause long-lasting depression of the ALDH activity in blood.[205]

An increased level of the lysosomal enzyme β-hexosaminidase has been observed in serum from alcoholics, and was proposed as a more sensitive marker of heavy drinking than GGT.[208,209] In a study on patients undergoing detoxification, the B-isoforms of the enzyme compared well with CDT in terms of sensitivity and disappearance rate from the circulation during the alcohol withdrawal phase.[210] However, the sensitivity is much lower when unselected populations are examined,[211] and it should be noted that serum β-hexosaminidase is elevated not only after alcohol abuse but also in non-alcoholic liver disease, diabetes, hypertension, and pregnancy,[212] implying a low specificity for alcohol.

As mentioned earlier in this chapter (Section 4.3.3), trace amounts (~0.1 mg/dL) of both methanol and ethanol are produced naturally in the body during the course of intermediary metabolism or via bacteria in the colon acting on dietary carbohydrates. When the blood-ethanol concentration exceeds 20 mg/dL, the hepatic Class I isozymes of ADH are saturated and therefore fully engaged in the clearance of ethanol from the body. Methanol, however, continues to be produced from endogenous substrates and its oxidation is hindered because of a competition with ethanol for available ADH enzymes. Furthermore, the concentration of methanol in blood increases during the drinking spree, owing to the congener content of the alcoholic beverages consumed. Recent studies have shown that this metabolic interaction between methanol and ethanol can help to distinguish between acute and chronic drinking practices.[213] After acute intake of ethanol, the BAC rarely exceeds 150 mg/dL and the corresponding blood methanol is generally below 0.5 mg/dL. However, if the BAC reaches 250 mg/dL, this suggests a period of continuous heavy drinking and the concentration of methanol now tends to increase in body fluids.[214] Depending on the intensity and duration of the drinking spree, as well as the methanol content of the drinks consumed,[81] the concentration of methanol in blood might increase appreciably and sometimes exceeds 1 mg/dL. Indeed, this concentration threshold (>1 mg/dL) is assumed to indicate continuous heavy drinking and therefore has been used as a marker of chronic alcoholism.[213] Because the metabolism of methanol is blocked after BAC exceeds 20 mg/dL (see Figure 4.3), the longer lasting the drinking spree so that BAC never drops below 20 mg/dL, the higher will be the steady-state concentration of methanol. The analysis of methanol as well as other biochemical markers of alcohol abuse has been used in Germany to decide on re-granting of driving licenses and whether a convicted drunk driver should receive treatment or punishment for this traffic crime.[213]

4.5 ROUTINE CLINICAL USE OF BIOCHEMICAL TESTS FOR EXCESSIVE DRINKING

4.5.1 Single Tests or Test Combinations?

As already indicated, most of the currently available standard laboratory tests lack sufficient sensitivity and/or specificity to warrant their use as sole evidence of heavy drinking. To increase diagnostic sensitivity, various combinations of markers have therefore been evaluated, such as two or more of GGT, MCV, AST, ALT, and CDT, as well as many others.[215–217] Even though combinations of markers tend to increase diagnostic sensitivity, at the same time this approach might reduce diagnostic specificity.[218–220] Furthermore, using multiple markers tends to complicate the interpretation of results and also increases the overall costs.

Nevertheless, a useful approach is to combine markers each independently associated with heavy drinking.[26] Whereas a strong association is usually obtained between liver function tests like GGT, AST, and ALT, this is not the case for GGT and CDT.[221–223] Rather, in one study, several of the highest CDT levels were observed in alcoholic subjects possessing normal or only moderately elevated GGT values, and vice versa.[222] This was further confirmed by another study in which a negative correlation between CDT and the severity of liver disease was reported.[224] It seems that the combined use of GGT and CDT significantly improves identification of heavy drinkers and those at risk of becoming dependent on alcohol.[219,222,223]

A combination of short-term and long-term markers of alcohol consumption has also been used successfully during outpatient treatment of alcohol-dependent subjects.[128,129] Finding an increased CDT suggests the person has returned to continuous heavy drinking after a period of abstinence, whereas normal CDT but instead a positive morning breath alcohol test or, preferably, other more sensitive biomarkers of acute alcohol intake, such as urinary 5HTOL, EtG, or EtS, will identify single lapses.

4.5.2 Screening for Excessive Drinking in Unselected Populations

Many biochemical screening markers used for the early detection of harmful consumption of alcohol give excellent results (i.e., high sensitivity and specificity) when studies are carried out on selected populations, e.g., alcoholics undergoing detoxification as compared to moderate drinkers or teetotalers. By contrast, in studies on single individuals, and when screening people from the general population, most tests are less satisfactory, mainly because of a considerable overlap between the values obtained for heavy and moderate drinkers. Furthermore, because certain medical and environmental conditions may influence the test results, false positives are not uncommon. An example illustrates this point. Assuming that the prevalence of excessive drinking is 10%, a marker with 90% sensitivity and specificity will correctly identify 9 of 10 heavy drinkers in a study population of 100, but, at the same time, incorrectly identify 9 of 90 healthy subjects as having drinking problems. Thus, the chance of a correct classification under these conditions is only 50% (the positive predictive value). For most of the currently available laboratory markers of alcohol abuse, a single abnormal test result may thus be difficult to interpret in an unequivocal way, unless confirmed in a repeated test or by complementary testing to exclude other potential causes.

Another important issue is the time delay between drinking and sampling. Because different alcohol biomarkers have different life spans, the time since the last drink should always be considered when evaluating the sensitivity of a test. The most widely used markers GGT and CDT, for example, have biological half-lives of ~2 to 3 weeks, so the specimen should preferably be collected no later than 1 week after admission to hospital or during alcohol withdrawal; otherwise the sensitivity of the test will be reduced considerably.

4.5.3 Treatment Follow-Up of Alcohol-Dependent Patients

Monitoring changes over time in a number of biomarkers of excessive drinking, such as CDT and GGT, during a 2- to 4-week period of alcohol withdrawal (e.g., during hospitalization or other inpatient treatment, or treatment with disulfiram), allows the most sensitive single marker to be identified.[222,225] If values normalize on withdrawal, this confirms that alcohol was most likely the cause of the abnormal test results. After discharge from the hospital, monitoring excessive drinking should continue on a routine basis in connection with return visits to the clinic. This facilitates early identification of relapse,[226] when combined with an early morning breath alcohol test or a more sensitive biomarker of acute drinking such as 5HTOL, EtG, and EtS.[26] The optimal testing frequency depends in part on the life span of the marker in question.[227] Giving feedback to patients about the results of the tests, for example, by presenting these graphically, can be very informative and may also improve self-report and treatment outcome.[228,229] However, if this strategy is used, it is imperative that the results are reliable (i.e., using highly specific tests); otherwise patients may become demoralized and lose faith in the use of laboratory markers.

The use of repeated testing also makes it possible to use individualized instead of population-based reference limits for each biomarker, thereby improving considerably the reliability of results.[129,149] The recommended cutoff limits between normal and abnormal are usually based on a Gaussian distribution in healthy controls. However, because of inter-individual variations, some subjects with very low baseline values probably need to drink much larger amounts of alcohol than those with considerably higher baseline values in order to exceed the critical threshold limit. Therefore, by introducing individualized cutoffs, detection of relapse drinking can be significantly improved.[129,230]

4.6 TRAIT MARKERS OF ALCOHOL DEPENDENCE

Eugenic studies conducted during the late 1800s demonstrated that excessive drinking, drunkenness, and alcohol dependence run in families. Unequivocal evidence for a genetic component in alcoholism and addiction came from the widely publicized adoption and twin studies.[231,232] Later research dealing with the inheritance of alcohol dependence led to the definitions of *type 1* and *type 2* subtypes, which have become widely accepted.[233] It seems that some people are more predisposed to becoming dependent on alcohol when they start to drink regularly, and, for this reason, efforts have been made to develop trait markers of alcohol dependence.[234–236]

Examples of such genetic or trait markers of a vulnerability to develop alcohol dependence include various neurotransmitter systems, such as the dopamine D_2 receptor gene, the activity of monoamine oxidase (MAO), and adenylyl cyclase (AC) enzymes and the serotonin transporter (5HTT), as well as the neuropeptides (e.g., neuropeptide Y, NPY), which are all involved in various aspects of compulsive, impulsive, and addictive behavior, including pleasure seeking and reward areas of the brain.[236,237] A significantly lower activity of MAO in blood platelets of alcoholics compared to controls was reported,[238] although concerns have been raised about the exact mechanism causing this difference.[239] Tobacco use, for example, which is common in alcoholics, can also lower MAO activity.[240]

Hitherto, the only clear-cut evidence linking alcohol use and abuse to genetics is the polymorphisms observed for ADH and ALDH enzymes in Japanese and some other Asian populations.[241] The presence of an inactive mitochondrial ALDH2 isozyme makes these individuals hypersensitive to acetaldehyde, which reaches an abnormally high concentration in blood even after drinking small amounts of alcohol. This inborn Antabuse-like reaction creates an aversion to alcohol, which influences drinking behavior and decreases the risk of becoming dependent on alcohol and developing alcoholic liver disease.[242]

4.7 CONCLUSIONS

Alcohol abuse and dependence are major risk factors for serious ill-health, social, and economic problems, and excessive drinking is a common cause of injuries and premature death.[243] Early identification of those who overindulge in drinking enhances the possibility of finding a successful treatment. However, obtaining accurate information about a person's drinking habits represents one of the major problems in the detection of excessive alcohol consumption. Self-reports of drinking practices during interview or in diagnostic questionnaires are widely used by clinicians for this purpose. Experience has shown that people with alcohol-related problems may deliberately deny or underreport the actual amounts they consume, at least in the early stages of misuse. For this reason, erroneous classification and underdiagnosis of alcohol abuse are fairly common.[146,244,245] To rectify this problem, efforts have been made to discover biochemical and hematological abnormalities associated with excessive drinking. These markers need to be sufficiently sensitive and specific to identify heavy drinkers, even when the person refrains from drinking prior to visiting a physician.

The perfect biochemical test should be specific for alcohol and also exhibit high sensitivity to hazardous drinking habits. Furthermore, the test should be inexpensive and yield rapid results with easily available specimens such as urine, saliva, or blood. Measuring ethanol in body fluids or breath is of course the most highly specific alcohol test, but, because ethanol is eliminated fairly rapidly from the body, the sensitivity of this test is low. Most of the currently available laboratory markers perform well when selected high-risk populations are being compared, but they are less satisfactory in randomly selected individuals. An alcohol breath test is simple to perform but this needs to be combined with a battery of more sensitive biochemical indicators, as discussed in this chapter, such as GGT and CDT. When the aim is to test for acute alcohol intake or relapse, the recommended markers besides ethanol are 5HTOL, EtG, or EtS.

Alcohol is a legal, socially accepted, recreational drug, although the patterns of alcohol consumption, in terms of quantity and the choice of beverage, differ widely between individuals and nations. Furthermore, a pattern of drinking that might be harmless to one individual might damage health in another, owing to various genetic influences and risk of becoming dependent on alcohol (see Chapter 1). Over the past 10 to 15 years, considerable interest has developed in using biochemical markers to document alcohol use and abuse in a more objective way, compared with clinical interviews and screening questionnaires about the negative consequences and experiences after a drinking spree. Identifying people at greatest risk for developing alcohol problems is a challenge, and positive findings will ensure adequate and early treatment and rehabilitation programs. In this connection, biochemical markers have emerged as essential tools to verify abstinence and to detect relapse drinking in patients undergoing treatment for alcoholism.

REFERENCES

1. Doll, R., The benefit of alcohol in moderation. *Drug Alcohol Rev.,* 17, 353, 1998.
2. Hanson, G.R. and Li, T.K., Public health implications of excessive alcohol consumption. *J. Am. Med. Assoc.,* 289, 1031, 2003.
3. Room, R., Babor, T., and Rehm, J., Alcohol and public health. *Lancet,* 365, 519, 2005.
4. Cherpitel, C.J., Drinking patterns and problems and drinking in the injury event: an analysis of emergency room patients by ethnicity. *Drug Alcohol Rev.,* 17, 423, 1998.
5. Connor, J., Norton, R., Ameratunga, S., and Jackson, R., The contribution of alcohol to serious car crash injuries. *Epidemiology,* 15, 337, 2004.
6. Richardson, A. and Budd, T., Young adults, alcohol, crime and disorder. *Crim. Behav. Ment. Health,* 13, 5, 2003.
7. Brewer, R.D. and Swahn, M.H., Binge drinking and violence. *J. Am. Med. Assoc.,* 294, 616, 2005.
8. Serdula, M.K., Brewer, R.D., Gillespie, C., Denny, C.H., and Mokdad, A., Trends in alcohol use and binge drinking, 1985–1999: results of a multi-state survey. *Am. J. Prev. Med.,* 26, 294, 2004.
9. Borges, G., Cherpitel, C., and Mittleman, M., Risk of injury after alcohol consumption: a case-crossover study in the emergency department. *Soc. Sci. Med.,* 58, 1191, 2004.
10. Hall, W., British drinking: a suitable case for treatment? *Br. Med. J.,* 331, 527, 2005.
11. Meyerhoff, D.J., Bode, C., Nixon, S.J., de Bruin, E.A., Bode, J.C., and Seitz, H.K., Health risks of chronic moderate and heavy alcohol consumption: how much is too much? *Alcohol Clin. Exp. Res.,* 29, 1334, 2005.
12. Jellinek, E.M., *The Disease Concept of Alcoholism*, Hillhouse Press, New Haven, CT, 1960.
13. Fuller, R.K., Lee, K.K., and Gordis, E., Validity of self-report in alcoholism research: results of a Veterans Administration cooperative study. *Alcohol Clin. Exp. Res.,* 12, 201, 1988.
14. Ness, D.E. and Ende, J., Denial in the medical interview: recognition and management. *J. Am. Med. Assoc.,* 272, 1777, 1994.
15. Wilson, R.S., *Diagnosis of Alcohol Abuse*, CRC Press, Boca Raton, FL, 1989.
16. Mayfield, D., McLeod, G., and Hall, P., The CAGE questionnaire: validation of a new alcoholism screening instrument. *Am. J. Psychiatry,* 131, 1121, 1974.

17. Selzer, M.L., The Michigan Alcoholism Screening Test: the quest for a new diagnostic instrument. *Am. J. Psychiatry,* 127, 89, 1981.
18. Mihas, A.A. and Tavassoli, M., Laboratory markers of ethanol intake and abuse: a critical appraisal. *Am. J. Med. Sci.,* 303, 415, 1992.
19. Goldberg, D.M. and Kapur, B.M., Enzymes and circulating proteins as markers of alcohol abuse. *Clin. Chim. Acta,* 226, 191, 1994.
20. Conigrave, K.M., Saunders, J.B., and Whitfield, J.B., Diagnostic tests for alcohol consumption. *Alcohol Alcohol.,* 30, 13, 1995.
21. Helander, A., Biological markers in alcoholism. *J. Neural Transm.,* 66(Suppl.), 15, 2003.
22. Salaspuro, M., Characteristics of laboratory markers in alcohol-related organ damage. *Scand. J. Gastroenterol.,* 24, 769, 1989.
23. Rosman, A.S. and Lieber, C.S., Diagnostic utility of laboratory tests in alcoholic liver disease. *Clin. Chem.,* 40, 1641, 1994.
24. Wilson, R.S., Mohs, M.E., Eskelson, C., Sampliner, R.E., and Hartmann, B., Identification of alcohol abuse and alcoholism with biological parameters. *Alcohol Clin. Exp. Res.,* 10, 364, 1986.
25. Sharpe, P.C., Biochemical detection and monitoring of alcohol abuse and abstinence. *Ann. Clin. Biochem.,* 38, 652, 2001.
26. Helander, A., Biological markers of alcohol use and abuse in theory and practice, in *Alcohol in Health and Disease*, Agarwal, D.P. and Seitz, H.K., Eds., Marcel Dekker, New York, 2001, 177.
27. Hermansson, U., Helander, A., Huss, A., Brandt, L., and Rönnberg, S., The Alcohol Use Disorders Identification Test (AUDIT) and carbohydrate-deficient transferrin (CDT) in a routine workplace health examination. *Alcohol Clin. Exp. Res.,* 24, 180, 2000.
28. Hermansson, U., Knutsson, A., Brandt, L., Huss, A., Rönnberg, S., and Helander, A., Screening for high-risk and elevated alcohol consumption in day and shift workers by use of the AUDIT and CDT. *Occup. Med.* (London), 53, 518, 2003.
29. Helander, A., Beck, O., and Jones, A.W., Distinguishing ingested ethanol from microbial formation by analysis of urinary 5-hydroxytryptophol and 5-hydroxyindoleacetic acid. *J. Forensic Sci.,* 40, 95, 1995.
30. Helander, A. and Jones, A.W., 5-HTOL — a new biochemical alcohol marker with forensic applications. *Lakartidningen,* 99, 3950, 2002.
31. Bergström, J., Helander, A., and Jones, A.W., Ethyl glucuronide concentrations in two successive urinary voids from drinking drivers: relationship to creatinine content and blood and urine ethanol concentrations. *Forensic Sci. Int.,* 133, 86, 2003.
32. Jones, A.W., Eklund, A., and Helander, A., Misleading results of ethanol analysis in urine specimens from rape victims suffering from diabetes. *J. Clin. Forensic Med.,* 7, 144, 2000.
33. Balldin, J., Berglund, M., Borg, S., Månsson, M., Bendtsen, P., Franck, J., Gustafsson, L., Halldin, J., Nilsson, L.H., Stolt, G., and Willander, A., A 6-month controlled naltrexone study: combined effect with cognitive behavioral therapy in outpatient treatment of alcohol dependence. *Alcohol Clin. Exp. Res.,* 27, 1142, 2003.
34. Anton, R.F., Moak, D.H., Latham, P., Waid, L.R., Myrick, H., Voronin, K., Thevos, A., Wang, W., and Woolson, R., Naltrexone combined with either cognitive behavioral or motivational enhancement therapy for alcohol dependence. *J. Clin. Psychopharmacol.,* 25, 349, 2005.
35. Zweig, M.H. and Campbell, G., Receiver-operating characteristic (ROC) plots: a fundamental evaluation tool in clinical medicine. *Clin. Chem.,* 39, 561, 1993.
36. Henderson, A.R., Assessing test accuracy and its clinical consequences: a primer for receiver operating characteristic curve analysis. *Ann. Clin. Biochem.,* 30, 521, 1993.
37. Helander, A., Beck, O., and Jones, A.W., Laboratory testing for recent alcohol consumption: comparison of ethanol, methanol, and 5-hydroxytryptophol. *Clin. Chem.,* 42, 618, 1996.
38. Jones, A.W. and Helander, A., Disclosing recent drinking after alcohol has been cleared from the body. *J. Anal. Toxicol.,* 20, 141, 1996.
39. Helander, A., von Wachenfeldt, J., Hiltunen, A., Beck, O., Liljeberg, P., and Borg, S., Comparison of urinary 5-hydroxytryptophol, breath ethanol, and self-report for detection of recent alcohol use during outpatient treatment: a study on methadone patients. *Drug Alcohol Depend.,* 56, 33, 1999.
40. Jones, A.W., Measuring alcohol in blood and breath for forensic purposes: a historical review. *Forensic Sci. Rev.,* 8, 13, 1996.

41. Dubowski, K.M. and Caplan, Y.H., Alcohol testing in the workplace, in *Medicolegal Aspects of Alcohol*, Garriott, J.C., Ed., Lawyers & Judges, Tucson, 1996, 439.
42. Haeckel, R. and Hanecke, P., Application of saliva for drug monitoring. An *in vivo* model for transmembrane transport. *Eur. J. Clin. Chem. Clin. Biochem.*, 34, 171, 1996.
43. Toennes, S.W., Kauert, G.F., Steinmeyer, S., and Moeller, M.R., Driving under the influence of drugs — evaluation of analytical data of drugs in oral fluid, serum and urine, and correlation with impairment symptoms. *Forensic Sci. Int.*, 152, 149, 2005.
44. Jones, A.W., Measuring ethanol in saliva with the QED enzymatic test device: comparison of results with blood and breath-alcohol concentration. *J. Anal. Toxicol.*, 19, 169, 1995.
45. Degutis, L.C., Rabinovici, R., Sabbaj, A., Mascia, R., and D'Onofrio, G., The saliva strip test is an accurate method to determine blood alcohol concentration in trauma patients. *Acad. Emerg. Med.*, 11, 885, 2004.
46. Jones, A.W., Norberg, A., and Hahn, R.G., Concentration-time profiles of ethanol in arterial and venous blood and end-expired breath during and after intravenous infusion. *J. Forensic Sci.*, 42, 1088, 1997.
47. Jones, A.W., Reference limits for urine/blood ratios of ethanol in two successive voids from drinking drivers. *J. Anal. Toxicol.*, 26, 333, 2002.
48. Bendtsen, P., Jones, A.W., and Helander, A., Urinary excretion of methanol and 5-hydroxytryptophol as biochemical markers of recent drinking in the hangover state. *Alcohol Alcohol.*, 33, 431, 1998.
49. Pawan, G.L.S. and Grice, K., Distribution of alcohol in urine and sweat after drinking. *Lancet*, 2, 1016, 1968.
50. Swift, R.M., Martin, C.S., Swette, L., LaConti, A., and Kackley, N., Studies on a wearable, electronic, transdermal alcohol sensor. *Alcohol Clin. Exp. Res.*, 16, 721, 1992.
51. Phillips, M., Greenberg, J., and Andrzejewski, J., Evaluation of the Alcopatch, a transdermal dosimeter for monitoring alcohol consumption. *Alcohol Clin. Exp. Res.*, 19, 1547, 1995.
52. Cone, E.J., Hillsgrove, M.J., Jenkins, A.J., Keenan, R.M., and Darwin, W.D., Sweat testing for heroin, cocaine, and metabolites. *J. Anal. Toxicol.*, 18, 298, 1994.
53. Huestis, M.A., Cone, E.J., Wong, C.J., Umbricht, A., and Preston, K.L., Monitoring opiate use in substance abuse treatment patients with sweat and urine drug testing. *J. Anal. Toxicol.*, 24, 509, 2000.
54. Phillips, M., Sweat patch test for alcohol consumption: rapid assay with an electrochemical detector. *Alcohol Clin. Exp. Res.*, 6, 532, 1982.
55. Swift, R., Davidson, D., and Fitz, E., Transdermal alcohol detection with a new miniaturized sensor, the miniTAS. *Alcohol Clin. Exp. Res.*, 20, 45A, 1996.
56. Swift, R., Direct measurement of alcohol and its metabolites. *Addiction*, 98(Suppl. 2), 73, 2003.
57. Mani, J.C., Pietruszko, R., and Theorell, H., Methanol activity of alcohol dehydrogenase from human liver, horse liver, and yeast. *Arch. Biochem. Biophys.*, 140, 52, 1970.
58. Eriksson, C.J.P. and Fukunaga, T., Human blood acetaldehyde (update 1992). *Alcohol Alcohol.*, 2(Suppl.), 9, 1993.
59. Helander, A., Löwenmo, C., and Johansson, M., Distribution of acetaldehyde in human blood: effects of ethanol and treatment with disulfiram. *Alcohol Alcohol.*, 28, 461, 1993.
60. Jones, A.W., Measuring and reporting the concentration of acetaldehyde in human breath. *Alcohol Alcohol.*, 30, 271, 1995.
61. Lundquist, F., Production and utilization of free acetate in man. *Nature*, 193, 579, 1962.
62. Nuutinen, H., Lindros, K., Hekali, P., and Salaspuro, M., Elevated blood acetate as indicator of fast ethanol elimination in chronic alcoholics. *Alcohol*, 2, 623, 1985.
63. Korri, U.-M., Nuutinen, H., and Salaspuro, M., Increased blood acetate: a new laboratory marker of alcoholism and heavy drinking. *Alcohol Clin. Exp. Res.*, 9, 468, 1985.
64. Roine, R.P., Korri, U.-M., Ylikahri, R., Pentillä, A., Pikkarainen, J., and Salapuro, M., Increased serum acetate as a marker of problem drinking among drunken drivers. *Alcohol Alcohol.*, 23, 123, 1988.
65. Goldberger, B.A., Cone, E.J., and Kadehjian, L., Unsuspected ethanol ingestion through soft drinks and flavored beverages. *J. Anal. Toxicol.*, 20, 332, 1996.
66. Logan, B.K. and Distefano, S., Ethanol content of various foods and soft drinks and their potential for interference with a breath-alcohol test. *J. Anal. Toxicol.*, 22, 181, 1998.
67. Logan, B.K. and Jones, A.W., Endogenous ethanol "auto-brewery syndrome" as a drunk-driving defence challenge. *Med. Sci. Law*, 40, 206, 2000.

68. Ostrovsky, Y.M., Endogenous ethanol — its metabolic, behavioral and biomedical significance. *Alcohol,* 3, 239, 1986.
69. Krebs, H.A. and Perkins, J.R., The physiological role of liver alcohol dehydrogenase. *Biochem. J.,* 118, 635, 1970.
70. Sprung, R., Bonte, W., Rüdell, E., Domke, M., and Frauenrath, C., Zum Problem des endogenen Alkohols. *Blutalkohol,* 18, 65, 1981.
71. Haffner, H.-T., Graw, M., Besserer, K., Blicke, U., and Henssge, C., Endogenous methanol: variability in concentration and rate of production. Evidence of a deep compartment? *Forensic Sci. Int.,* 79, 145, 1996.
72. Jones, A.W., Elimination half-life of methanol during hangover. *Pharmacol. Toxicol.,* 60, 217, 1987.
73. Haffner, H.-T., Wehner, H.D., Scheytt, K.D., and Besserer, K., The elimination kinetics of methanol and the influence of ethanol. *Int. J. Leg. Med.,* 105, 111, 1992.
74. Helander, A. and Eriksson, C.J., Laboratory tests for acute alcohol consumption: results of the WHO/ISBRA Study on State and Trait Markers of Alcohol Use and Dependence. *Alcohol Clin. Exp. Res.,* 26, 1070, 2002.
75. Jacobsen, D., Jansen, H., Wiik-Larsen, E., Bredesen, J.-E., and Halvorsen, S., Studies on methanol poisoning. *Acta Med. Scand.,* 212, 5, 1982.
76. Prabhakaran, V., Ettler, H., and Mills, A., Methanol poisoning: two cases with similar plasma methanol concentrations but different outcomes. *Can. Med. Assoc. J.,* 148, 981, 1993.
77. Hantson, P., Wittebole, X., and Haufroid, V., Ethanol therapy for methanol poisoning: duration and problems. *Eur. J. Emerg. Med.,* 9, 278, 2002.
78. Jacobsen, D. and McMartin, K.E., Antidotes for methanol and ethylene glycol poisoning. *J. Toxicol. Clin. Toxicol.,* 35, 127, 1997.
79. Brent, J., McMartin, K., Phillips, S., Aaron, C., and Kulig, K., Fomepizole for the treatment of methanol poisoning. *N. Engl. J. Med.,* 344, 424, 2001.
80. McAnalley, B.H., Chemistry of alcoholic beverages, in *Medicolegal Aspects of Alcohol*, Garriott, J.C., Ed., Lawyers & Judges, Tucson, 1996, 1.
81. Gilg, T., Alkoholbedingte Fahruntuchtigkeit. *Rechtsmedizin,* 15, 97, 2005.
82. Majchrowicz, E. and Mendelson, J.H., Blood methanol concentrations during experimentally induced ethanol intoxication in alcoholics. *J. Pharmacol. Exp. Ther.,* 179, 293, 1971.
83. Iffland, R., New ways to use biochemical indicators of alcohol abuse to regrant licences in a fairer manner after drunken driving in Germany. *Alcohol Alcohol.,* 31, 619, 1996.
84. Haffner, H.T., Banger, M., Graw, M., Besserer, K., and Brink, T., The kinetics of methanol elimination in alcoholics and the influence of ethanol. *Forensic Sci. Int.,* 89, 129, 1997.
85. Graw, M., Haffner, H.T., Althaus, L., Besserer, K., and Voges, S., Invasion and distribution of methanol. *Arch. Toxicol.,* 74, 313, 2000.
86. Taucher, J., Lagg, A., Hansel, A., Vogel, W., and Lindinger, W., Methanol in human breath. *Alcohol Clin. Exp. Res.,* 19, 1147, 1995.
87. Lindinger, W., Taucher, J., Jordan, A., Hansel, A., and Vogel, W., Endogenous production of methanol after the consumption of fruit. *Alcohol Clin. Exp. Res.,* 21, 939, 1997.
88. Bachmann, C. and Bickel, M.H., History of drug metabolism: the first half of the 20th century. *Drug Metab. Rev.,* 16, 185, 1985.
89. Neubauer, O., Uber Glukuronsaure paarung bei Stoffen der Fettreihe. *Arch. Exp. Pathol. Pharmacol.,* 46, 133, 1901.
90. Kamil, I.A., Smith, J.N., and Williams, R.T., A new aspect of ethanol metabolism: isolation of ethyl glucuronide. *Biochem. J.,* 51, 32, 1952.
91. Jaakonmaki, P.I., Knox, K.L., Horning, E.C., and Horning, M.G., The characterization by gas-liquid chromatography of ethyl β-D-glucosiduronic acid as a metabolite of ethanol in rat and man. *Eur. J. Pharmacol.,* 1, 63, 1967.
92. Hanai, T., Chromatography and computational chemical analysis for drug discovery. *Curr. Med. Chem.,* 12, 501, 2005.
93. Dahl, H., Stephanson, N., Beck, O., and Helander, A., Comparison of urinary excretion characteristics of ethanol and ethyl glucuronide. *J. Anal. Toxicol.,* 26, 201, 2002.
94. Helander, A. and Beck, O., Mass spectrometric identification of ethyl sulfate as an ethanol metabolite in humans. *Clin. Chem.,* 50, 936, 2004.

95. Schmitt, G., Aderjan, R., Keller, T., and Wu, M., Ethyl glucuronide: an unusual ethanol metabolite in humans. Synthesis, analytical data, and determination in serum and urine. *J. Anal. Toxicol.,* 19, 91, 1995.
96. Schmitt, G., Droenner, P., Skopp, G., and Aderjan, R., Ethyl glucuronide concentration in serum of human volunteers, teetotalers, and suspected drinking drivers. *J. Forensic Sci.,* 42, 1099, 1997.
97. Helander, A. and Beck, O., Ethyl sulfate: a metabolite of ethanol in humans and a potential biomarker of acute alcohol intake. *J. Anal. Toxicol.,* 29, 270, 2005.
98. Sarkola, T., Dahl, H., Eriksson, C.J., and Helander, A., Urinary ethyl glucuronide and 5-hydroxytryptophol levels during repeated ethanol ingestion in healthy human subjects. *Alcohol Alcohol.,* 38, 347, 2003.
99. Skopp, G., Schmitt, G., Potsch, L., Dronner, P., Aderjan, R., and Mattern, R., Ethyl glucuronide in human hair. *Alcohol Alcohol.,* 35, 283, 2000.
100. Yegles, M., Labarthe, A., Auwarter, V., Hartwig, S., Vater, H., Wennig, R., and Pragst, F., Comparison of ethyl glucuronide and fatty acid ethyl ester concentrations in hair of alcoholics, social drinkers and teetotallers. *Forensic Sci. Int.,* 145, 167, 2004.
101. Stephanson, N., Dahl, H., Helander, A., and Beck, O., Direct quantification of ethyl glucuronide in clinical urine samples by liquid chromatography-mass spectrometry. *Ther. Drug Monit.,* 24, 645, 2002.
102. Nishikawa, M., Tsuchihashi, H., Miki, A., Katagi, M., Schmitt, G., Zimmer, H., Keller, T., and Aderjan, R., Determination of ethyl glucuronide, a minor metabolite of ethanol, in human serum by liquid chromatography-electrospray ionization mass spectrometry. *J. Chromatogr. B Biomed. Sci. Appl.,* 726, 105, 1999.
103. Seidl, S., Wurst, F.M., and Alt, A., Ethyl glucuronide — a biological marker for recent alcohol consumption. *Addict. Biol.,* 6, 205, 2001.
104. Helander, A. and Dahl, H., Urinary tract infection: a risk factor for false-negative urinary ethyl glucuronide but not ethyl sulfate in the detection of recent alcohol consumption. *Clin. Chem.,* 51, 1728, 2005.
105. Schneider, H. and Glatt, H., Sulpho-conjugation of ethanol in humans *in vivo* and by individual sulphotransferase forms *in vitro. Biochem. J.,* 383, 543, 2004.
106. Dresen, S., Weinmann, W., and Wurst, F.M., Forensic confirmatory analysis of ethyl sulfate — a new marker for alcohol consumption — by liquid-chromatography/electrospray ionization/tandem mass spectrometry. *J. Am. Soc. Mass. Spectrom.,* 15, 1644, 2004.
107. Deng, X.S., Bludeau, P., and Deitrich, R.A., Formation of ethyl nitrite *in vivo* after ethanol administration. *Alcohol,* 34, 217, 2004.
108. Lange, L.G. and Sobel, B.E., Myocardial metabolites of ethanol. *Circ. Res.,* 52, 479, 1983.
109. Laposata, M., Hasaba, A., Best, C.A., Yoerger, D.M., McQuillan, B.M., Salem, R.O., Refaai, M.A., and Soderberg, B.L., Fatty acid ethyl esters: recent observations. *Prostaglandins Leukot. Essent. Fatty Acids,* 67, 193, 2002.
110. Kaphalia, B.S., Cai, P., Khan, M.F., Okorodudu, A.O., and Ansari, G.A., Fatty acid ethyl esters: markers of alcohol abuse and alcoholism. *Alcohol,* 34, 151, 2004.
111. Soderberg, B.L., Salem, R.O., Best, C.A., Cluette-Brown, J.E., and Laposata, M., Fatty acid ethyl esters. Ethanol metabolites that reflect ethanol intake. *Am. J. Clin. Pathol.,* 119(Suppl.), S94, 2003.
112. Bisaga, A., Laposata, M., Xie, S., and Evans, S.M., Comparison of serum fatty acid ethyl esters and urinary 5-hydroxytryptophol as biochemical markers of recent ethanol consumption. *Alcohol Alcohol.,* 40, 214, 2005.
113. Borucki, K., Schreiner, R., Dierkes, J., Jachau, K., Krause, D., Westphal, S., Wurst, F.M., Luley, C., and Schmidt-Gayk, H., Detection of recent ethanol intake with new markers: comparison of fatty acid ethyl esters in serum and of ethyl glucuronide and the ratio of 5-hydroxytryptophol to 5-hydroxyindole acetic acid in urine. *Alcohol Clin. Exp. Res.,* 29, 781, 2005.
114. Nanji, A.A., Su, G.L., Laposata, M., and French, S.W., Pathogenesis of alcoholic liver disease — recent advances. *Alcohol Clin. Exp. Res.,* 26, 731, 2002.
115. Davis, V.E., Brown, H., Huff, J.A., and Cashaw, J.L., The alteration of serotonin metabolism to 5-hydroxytryptophol by ethanol ingestion in man. *J. Lab. Clin. Med.,* 69, 132, 1967.
116. Helander, A., Beck, O., Jacobsson, G., Löwenmo, C., and Wikström, T., Time course of ethanol-induced changes in serotonin metabolism. *Life Sci.,* 53, 847, 1993.
117. Lahti, R.A. and Majchrowicz, E., Ethanol and acetaldehyde effects on metabolism and binding of biogenic amines. *Q. J. Stud. Alc.,* 35, 1, 1974.

118. Feldstein, A. and Williamson, O., 5-Hydroxytryptamine metabolism in rat brain and liver homogenates. *Br. J. Pharmacol.*, 34, 38, 1968.
119. Svensson, S., Some, M., Lundsjö, A., Helander, A., Cronholm, T., and Höög, J.O., Activities of human alcohol dehydrogenases in the metabolic pathways of ethanol and serotonin. *Eur. J. Biochem.*, 262, 324, 1999.
120. Voltaire, A., Beck, O., and Borg, S., Urinary 5-hydroxytryptophol: a possible marker of recent alcohol consumption. *Alcohol Clin. Exp. Res.*, 16, 281, 1992.
121. Helander, A., Beck, O., and Borg, S., The use of 5-hydroxytryptophol as an alcohol intake marker. *Alcohol Alcohol.*, Suppl. 2, 497, 1994.
122. Beck, O. and Helander, A., 5-Hydroxytryptophol as a marker for recent alcohol intake. *Addiction*, 98(Suppl. 2), 63, 2003.
123. Helander, A., Wikström, T., Löwenmo, C., Jacobsson, G., and Beck, O., Urinary excretion of 5-hydroxyindole-3-acetic acid and 5-hydroxytryptophol after oral loading with serotonin. *Life Sci.*, 50, 1207, 1992.
124. Helander, A., Walzer, C., Beck, O., Balant, L., Borg, S., and Wartburg, J.-P.V., Influence of genetic variation in alcohol and aldehyde dehydrogenase on serotonin metabolism. *Life Sci.*, 55, 359, 1994.
125. Beck, O., Helander, A., Carlsson, S., and Borg, S., Changes in serotonin metabolism during treatment with the aldehyde dehydrogenase inhibitors disulfiram and cyanamide. *Pharmacol. Toxicol.*, 77, 323, 1995.
126. Johnson, R.D., Lewis, R.J., Canfield, D.V., and Blank, C.L., Accurate assignment of ethanol origin in postmortem urine: liquid chromatographic-mass spectrometric determination of serotonin metabolites. *J. Chromatogr. B Anal. Technol. Biomed. Life Sci.*, 805, 223, 2004.
127. Stephanson, N., Dahl, H., Helander, A., and Beck, O., Determination of urinary 5-hydroxytryptophol glucuronide by liquid chromatography-mass spectrometry. *J. Chromatogr. B Anal. Technol. Biomed. Life Sci.*, 816, 107, 2005.
128. Voltaire Carlsson, A., Hiltunen, A.J., Beck, O., Stibler, H., and Borg, S., Detection of relapses in alcohol-dependent patients: comparison of carbohydrate-deficient transferrin in serum, 5-hydroxytryptophol in urine, and self-reports. *Alcohol Clin. Exp. Res.*, 17, 703, 1993.
129. Borg, S., Helander, A., Voltaire Carlsson, A., and Högstrom Brandt, A.M., Detection of relapses in alcohol-dependent patients using carbohydrate-deficient transferrin: improvement with individualized reference levels during long-term monitoring. *Alcohol Clin. Exp. Res.*, 19, 961, 1995.
130. Spies, C.D., Herpell, J., Beck, O., Muller, C., Pragst, F., Borg, S., and Helander, A., The urinary ratio of 5-hydroxytryptophol to 5-hydroxyindole-3-acetic acid in surgical patients with chronic alcohol misuse. *Alcohol*, 17, 19, 1999.
131. Helander, A., Beck, O., and Jones, A.W., Urinary 5HTOL/5HIAA as biochemical marker of postmortem ethanol synthesis. *Lancet*, 340, 1159, 1992.
132. Hagan, R.L. and Helander, A., Urinary 5-hydroxytryptophol following acute ethanol consumption: clinical evaluation and potential aviation applications. *Aviat. Space Environ. Med.*, 68, 30, 1997.
133. Johnson, R.D., Lewis, R.J., Canfield, D.V., Dubowski, K.M., and Blank, C.L., Utilizing the urinary 5-HTOL/5-HIAA ratio to determine ethanol origin in civil aviation accident victims. *J. Forensic Sci.*, 50, 670, 2005.
134. Rosalki, S.B. and Rau, D., Serum gamma-glutamyl transpeptidase activity in alcoholism. *Clin. Chim. Acta*, 39, 41, 1972.
135. Rosalki, S.B., Gamma-glutamyl transpeptidase. *Adv. Clin. Chem.*, 17, 53, 1975.
136. Nemesánszky, E. and Lott, J.A., Gamma-glutamyltransferase and its isoenzymes: progress and problems. *Clin. Chem.*, 31, 797, 1985.
137. Shaw, S. and Lieber, C.S., Mechanism of increased gamma glutamyl transpeptidase after chronic alcohol consumption: hepatic microsomal induction rather than dietary imbalance. *Substance Alcohol Actions/Misuse*, 1, 423, 1980.
138. Wu, A., Slavin, G., and Levi, A.J., Elevated serum gamma-glutamyl-transferase (transpeptidase) and histological liver damage in alcoholism. *Am. J. Gastroenterol.*, 65, 318, 1976.
139. Salaspuro, M., Use of enzymes for the diagnosis of alcohol related organ damage. *Enzyme*, 37, 87, 1987.
140. Nilssen, O. and Forde, O.H., Seven-year longitudinal population study of change in gamma-glutamyltransferase: the Tromsø study. *Am. J. Epidemiol.*, 139, 787, 1994.
141. Penn, R. and Worthington, L.J., Is serum gamma-glutamyltransferase a misleading test? *Br. Med. J.*, 286, 531, 1983.

142. Nalpas, R., Vassault, A., Charpin, S., Lacour, B., and Berthelot, P., Serum mitochondrial aspartate aminotransferase as a marker of chronic alcoholism: diagnostic value and interpretation in a liver unit. *Hepatology,* 6, 608, 1986.
143. Gluud, C., Andersen, I., Dietrichson, O., Gluud, B., Jacobsen, A., and Juhl, E., Gamma-glutamyltransferase, aspartate aminotransferase and alkaline phosphatase as marker of alcohol consumption in outpatient alcoholics. *Eur. J. Clin. Invest.,* 11, 171, 1981.
144. Nalpas, B., Vassault, A., Le Guillou, A., Lesgourgues, B., Ferry, N., Lacour, B., and Berthelot, P., Serum activity of mitochondrial asparate aminotransferase: a sensitive marker of alcoholism with or without alcoholic hepatitis. *Hepatology,* 5, 893, 1984.
145. Nalpas, R., Poupon, R.E., Vassault, A., Hauzanneau, P., Sage, Y., Schellenberg, F., Lacour, B., and Berthelot, P., Evaluations of mAST/tAST ratio as a marker of alcohol misuse in a non-selected population. *Alcohol Alcohol.,* 24, 415, 1989.
146. Nilssen, O., Huseby, N.E., Hoyer, G., Brenn, T., Schirmer, H., and Forde, O.H., New alcohol markers — how useful are they in population studies: the Svalbard Study 1988-89. *Alcohol Clin. Exp. Res.,* 16, 82, 1992.
147. Schiele, F., Artur, Y., Varasteh, A., Wellman, M., and Siest, G., Serum mitochondrial aspartate aminotransferase activity: not useful as marker of excessive alcohol consumption in an unselected population. *Clin. Chem.,* 35, 926, 1989.
148. Sillanaukee, P., Seppä, K., Lof, K., and Koivula, T., CDT by anion-exchange chromatography followed by RIA as a marker of heavy drinking among men. *Alcohol Clin. Exp. Res.,* 17, 230, 1993.
149. Irwin, M., Baird, S., Smith, T.L., and Schuckit, M., Use of laboratory tests to monitor heavy drinking by alcoholic men discharged from a treatment program. *Am. J. Psychiatry,* 145, 595, 1988.
150. Wu, A., Chanarin, I., and Levi, A.J., Macrocytosis in chronic alcoholism. *Lancet,* 1, 829, 1974.
151. Chick, J., Kreitman, N., and Plant, M., Mean cell volume and gamma-glutamyltranspeptidase as markers of drinking in working men. *Lancet,* 1, 1249, 1981.
152. Stimmel, B., Kurtz, D., Jackson, G., and Gilbert, H.S., Failure of mean red cell volume to serve as a biologic marker for alcoholism in narcotic dependence. *Am. J. Med.,* 74, 369, 1983.
153. Whitehead, T.P., Robinson, D., Allaway, S.L., and Hale, A.C., The effects of cigarette smoking and alcohol consumption on blood haemoglobin, erythrocytes and leucocytes: a dose related study on male subjects. *Clin. Lab. Haematol.,* 17, 131, 1995.
154. DiFranza, J.R. and Guerrera, M.P., Alcoholism and smoking. *J. Stud. Alcohol,* 51, 130, 1990.
155. Behrens, U.J., Worner, T.M., Braly, L.F., Schaffner, F., and Lieber, C.S., Carbohydrate-deficient transferrin, a marker for chronic alcohol consumption in different ethnic populations. *Alcohol Clin. Exp. Res.,* 12, 427, 1988.
156. Bell, H., Tallaksen, C.M., Try, K., and Haug, E., Carbohydrate-deficient transferrin and other markers of high alcohol consumption: a study of 502 patients admitted consecutively to a medical department. *Alcohol Clin. Exp. Res.,* 18, 1103, 1994.
157. Salaspuro, M., Carbohydrate-deficient transferrin as compared to other markers of alcoholism: a systematic review. *Alcohol,* 19, 261, 1999.
158. Stibler, H., Carbohydrate-deficient transferrin in serum: a new marker of potentially harmful alcohol consumption reviewed. *Clin. Chem.,* 37, 2029, 1991.
159. Jeppsson, J.O., Kristensson, H., and Fimiani, C., Carbohydrate-deficient transferrin quantified by HPLC to determine heavy consumption of alcohol. *Clin. Chem.,* 39, 2115, 1993.
160. de Jong, G., van Dijk, J.P., and van Eijk, H.G., The biology of transferrin. *Clin. Chim. Acta,* 190, 1, 1990.
161. Arndt, T., Carbohydrate-deficient transferrin as a marker of chronic alcohol abuse: a critical review of preanalysis, analysis, and interpretation. *Clin. Chem.,* 47, 13, 2001.
162. Helander, A., Eriksson, G., Stibler, H., and Jeppsson, J.O., Interference of transferrin isoform types with carbohydrate-deficient transferrin quantification in the identification of alcohol abuse. *Clin. Chem.,* 47, 1225, 2001.
163. Landberg, E., Påhlsson, P., Lundblad, A., Arnetorp, A., and Jeppsson, J.-O., Carbohydrate composition of serum transferrin isoforms from patients with high alcohol consumption. *Biochem. Biophys. Res. Commun.,* 210, 267, 1995.
164. Flahaut, C., Michalski, J.C., Danel, T., Humbert, M.H., and Klein, A., The effects of ethanol on the glycosylation of human transferrin. *Glycobiology,* 13, 191, 2003.

165. Peter, J., Unverzagt, C., Engel, W. D., Renauer, D., Seidel, C., and Hösel, W., Identification of carbohydrate deficient transferrin forms by MALDI-TOF mass spectrometry and lectin ELISA. *Biochim. Biophys. Acta,* 1380, 93, 1998.

166. Stibler, H. and Borg, S., Glycoprotein glycosyltransferase activities in serum in alcohol-abusing patients and healthy controls. *Scand. J. Clin. Lab. Invest.,* 51, 43, 1991.

167. Lieber, C.S., Carbohydrate deficient transferrin in alcoholic liver disease: mechanisms and clinical implications. *Alcohol,* 19, 249, 1999.

168. Sillanaukee, P., Strid, N., Allen, J.P., and Litten, R.Z., Possible reasons why heavy drinking increases carbohydrate-deficient transferrin. *Alcohol Clin. Exp. Res.,* 25, 34, 2001.

169. Allen, J.P., Litten, R.Z., Anton, R.F., and Cross, G.M., Carbohydrate-deficient transferrin as a measure of immoderate drinking: remaining issues. *Alcohol Clin. Exp. Res.,* 18, 799, 1994.

170. Conigrave, K.M., Degenhardt, L.J., Whitfield, J.B., Saunders, J.B., Helander, A., and Tabakoff, B., CDT, GGT, and AST as markers of alcohol use: the WHO/ISBRA collaborative project. *Alcohol Clin. Exp. Res.,* 26, 332, 2002.

171. Salmela, K.S., Laitinen, K., Nyström, M., and Salaspuro, M., Carbohydrate-deficient transferrin during 3 weeks' heavy alcohol consumption. *Alcohol Clin. Exp. Res.,* 18, 228, 1994.

172. Lesch, O.M., Walter, H., Antal, J., Heggli, D.E., Kovacz, A., Leitner, A., Neumeister, A., Stumpf, I., Sundrehagen, E., and Kasper, S., Carbohydrate-deficient transferrin as a marker of alcohol intake: a study with healthy subjects. *Alcohol Alcohol.,* 31, 265, 1996.

173. Kamboh, M.I. and Ferrell, R.E., Human transferrin polymorphism. *Hum. Hered.,* 37, 65, 1987.

174. Stibler, H., Borg, S., and Beckman, G., Transferrin phenotype and level of carbohydrate-deficient transferrin in healthy individuals. *Alcohol Clin. Exp. Res.,* 12, 450, 1988.

175. Bean, P. and Peter, J.B., Allelic D variants of transferrin in evaluation of alcohol abuse: differential diagnosis by isoelectric focusing-immunoblotting-laser densitometry. *Clin. Chem.,* 40, 2078, 1994.

176. Marquardt, T. and Denecke, J., Congenital disorders of glycosylation: review of their molecular bases, clinical presentations and specific therapies. *Eur. J. Pediatr.,* 162, 359, 2003.

177. Jaeken, J., Komrower Lecture. Congenital disorders of glycosylation (CDG): it's all in it! *J. Inherit. Metab. Dis.,* 26, 99, 2003.

178. Sorvajärvi, K., Blake, J.E., Israel, Y., and Niemelä, O., Sensitivity and specificity of carbohydrate-deficient transferrin as a marker of alcohol abuse are significantly influenced by alterations in serum transferrin: comparison of two methods. *Alcohol Clin. Exp. Res.,* 20, 449, 1996.

179. Helander, A., Absolute or relative measurement of carbohydrate-deficient transferrin in serum? Experiences with three immunological assays. *Clin. Chem.,* 45, 131, 1999.

180. Keating, J., Cheung, C., Peters, T.J., and Sherwood, R.A., Carbohydrate deficient transferrin in the assessment of alcohol misuse: absolute or relative measurements? A comparison of two methods with regard to total transferrin concentration. *Clin. Chim. Acta,* 272, 159, 1998.

181. Stibler, H., Borg, S., and Joustra, M., A modified method for the assay of carbohydrate-deficient transferrin (CDT) in serum. *Alcohol Alcohol.,* Suppl. 1, 451, 1991.

182. Helander, A., Husa, A., and Jeppsson, J.O., Improved HPLC method for carbohydrate-deficient transferrin in serum. *Clin. Chem.,* 49, 1881, 2003.

183. Helander, A., Wielders, J.P., te Stroet, R., and Bergström, J.P., Comparison of HPLC and capillary electrophoresis for confirmatory testing of the alcohol misuse marker carbohydrate-deficient transferrin. *Clin. Chem.,* 51, 1528, 2005.

184. Helander, A., Dahl, H., Swanson, I., and Bergström, J., Evaluation of Dade Behring N Latex CDT: a novel homogenous immunoassay for carbohydrate-deficient transferrin. *Alcohol Clin. Exp. Res.,* 28(Suppl.), 33A, 2004.

185. Kraul, D., Hackler, R., and Althaus, H., A novel particle-enhanced assay for the immuno-nephelometric determination of carbohydrate-deficient transferrin. *Alcohol Clin. Exp. Res.,* 28(Suppl.), 34A, 2004.

186. Alling, C., Gustavsson, L., and Änggård, E., An abnormal phospholipid in rat organs after ethanol treatment. *FEBS Lett.,* 152, 24, 1983.

187. Mueller, G.C., Fleming, M.F., LeMahieu, M.A., Lybrand, G.S., and Barry, K.J., Synthesis of phosphatidylethanol — a potential marker for adult males at risk for alcoholism. *Proc. Natl. Acad. Sci. U.S.A.,* 85, 9778, 1988.

188. Hansson, P., Caron, M., Johnson, G., Gustavsson, L., and Alling, C., Blood phosphatidylethanol as a marker of alcohol abuse: levels in alcoholic males during withdrawal. *Alcohol Clin. Exp. Res.,* 21, 108, 1997.

189. Varga, A., Hansson, P., Johnson, G., and Alling, C., Normalization rate and cellular localization of phosphatidylethanol in whole blood from chronic alcoholics. *Clin. Chim. Acta,* 299, 141, 2000.
190. Varga, A., Hansson, P., Lundqvist, C., and Alling, C., Phosphatidylethanol in blood as a marker of ethanol consumption in healthy volunteers: comparison with other markers. *Alcohol Clin. Exp. Res.,* 22, 1832, 1998.
191. Tolonen, A., Lehto, T.M., Hannuksela, M.L., and Savolainen, M.J., A method for determination of phosphatidylethanol from high density lipoproteins by reversed-phase HPLC with TOF-MS detection. *Anal. Biochem.,* 341, 83, 2005.
192. Aradottir, S., Seidl, S., Wurst, F.M., Jonsson, B.A., and Alling, C., Phosphatidylethanol in human organs and blood: a study on autopsy material and influences by storage conditions. *Alcohol Clin. Exp. Res.,* 28, 1718, 2004.
193. Stevens, V.J., Fantl, W.J., Newman, C.B., Sims, R.V., Cerami, A., and Peterson, C.M., Acetaldehyde adducts with hemoglobin. *J. Clin. Invest.,* 67, 361, 1981.
194. Tuma, D.J. and Casey, C.A., Dangerous byproducts of alcohol breakdown — focus on adducts. *Alcohol Res. Health,* 27, 285, 2003.
195. Niemelä, O. and Parkkila, S., Alcoholic macrocytosis — is there a role for acetaldehyde and adducts? *Addict. Biol.,* 9, 3, 2004.
196. Sorrell, M.F. and Tuma, D.J., Hypothesis: alcoholic liver injury and the covalent binding of acetaldehyde. *Alcohol Clin. Exp. Res.,* 9, 306, 1985.
197. Niemelä, O., Distribution of ethanol-induced protein adducts in vivo: relationship to tissue injury. *Free Radical Biol. Med.,* 31, 1533, 2001.
198. Lieber, C.S., Alcoholic fatty liver: its pathogenesis and mechanism of progression to inflammation and fibrosis. *Alcohol,* 34, 9, 2004.
199. Hoerner, M., Behrens, U.J., Worner, T., and Lieber, C.S., Humoral immune response to acetaldehyde adducts in alcoholic patients. *Res. Commun. Chem. Pathol. Pharmacol.,* 54, 3, 1986.
200. Peterson, K.P., Bowers, C., and Peterson, C.M., Prevalence of ethanol consumption may be higher in women than men in a university health service population as determined by a biochemical marker: whole blood-associated acetaldehyde above the 99th percentile for teetotalers. *J. Addict. Dis.,* 17, 13, 1998.
201. Itälä, L., Seppä, K., Turpeinen, U., and Sillanaukee, P., Separation of hemoglobin acetaldehyde adducts by high-performance liquid chromatography-cation-exchange chromatography. *Anal. Biochem.,* 224, 323, 1995.
202. Viitala, K., Israel, Y., Blake, J.E., and Niemelä, O., Serum IgA, IgG, and IgM antibodies directed against acetaldehyde-derived epitopes: relationship to liver disease severity and alcohol consumption. *Hepatology,* 25, 1418, 1997.
203. Sillanaukee, P., Seppä, K., Koivula, T., Israel, Y., and Niemelä, O., Acetaldehyde-modified hemoglobin as a marker of alcohol consumption: comparison of two new methods. *J. Lab. Clin. Med.,* 120, 42, 1992.
204. Halvorsen, M.R., Campbell, J.L., Sprague, G., Slater, K., Noffsinger, J.K., and Peterson, C.M., Comparative evaluation of the clinical utility of three markers of ethanol intake: the effect of gender. *Alcohol Clin. Exp. Res.,* 17, 225, 1993.
205. Helander, A., Aldehyde dehydrogenase in blood: distribution, characteristics and possible use as marker of alcohol misuse. *Alcohol Alcohol.,* 28, 135, 1993.
206. Johnson, R.D., Bahnisch, J., Stewart, B., Shearman, D.J.C., and Edwards, J.B., Optimized spectrophotometric determination of aldehyde dehydrogenase activity in erythrocytes. *Clin. Chem.,* 38, 584, 1992.
207. Hansell, N.K., Pang, D., Heath, A.C., Martin, N.G., and Whitfield, J.B., Erythrocyte aldehyde dehydrogenase activity: lack of association with alcohol use and dependence or alcohol reactions in Australian twins. *Alcohol Alcohol.,* 40, 343, 2005.
208. Kärkkäinen, P., Poikolainen, K., and Salaspuro, M., Serum β-hexosaminidase as a marker of heavy drinking. *Alcohol Clin. Exp. Res.,* 14, 187, 1990.
209. Hultberg, B., Isaksson, A., Berglund, M., and Moberg, A.-L., Serum β-hexosaminidase isoenzyme: a sensitive marker for alcohol abuse. *Alcohol Clin. Exp. Res.,* 15, 549, 1991.
210. Hultberg, B., Isaksson, A., Berglund, M., and Alling, C., Increases and time-course variations in beta-hexosaminidase isoenzyme B and carbohydrate-deficient transferrin in serum from alcoholics are similar. *Alcohol Clin. Exp. Res.,* 19, 452, 1995.

211. Nyström, M., Peräsalo, J., and Salaspuro, M., Serum β-hexosaminidase in young university students. *Alcohol Clin. Exp. Res.*, 15, 877, 1991.
212. Hultberg, B. and Isaksson, A., Isoenzyme pattern of serum β-hexosaminidase in liver disease, alcohol intoxication and pregnancy. *Enzyme*, 30, 166, 1983.
213. Iffland, R. and Grassnack, F., Untersuchung zum CDT und anderen Indikatoren für Alkoholprobleme im Blut alkoholauffälliger Pkw-Fahrer. *Blutalkohol*, 32, 26, 1995.
214. Haffner, H.-T., Batra, A., Wehner, H.D., Besserer, K., and Mann, K., Methanolspiegel und Methanolelimination bei Alkoholikern. *Blutalkohol*, 30, 52, 1993.
215. Hollstedt, C. and Dahlgren, L., Peripheral markers in the female "hidden alcoholic." *Acta Psychiatr. Scand.*, 75, 591, 1987.
216. Harasymiw, J. and Bean, P., The combined use of the early detection of alcohol consumption (EDAC) test and carbohydrate-deficient transferrin to identify heavy drinking behaviour in males. *Alcohol Alcohol.*, 36, 349, 2001.
217. Javors, M.A. and Johnson, B.A., Current status of carbohydrate deficient transferrin, total serum sialic acid, sialic acid index of apolipoprotein J and serum beta-hexosaminidase as markers for alcohol consumption. *Addiction*, 98(Suppl. 2), 45, 2003.
218. Sillanaukee, P., Aalto, M., and Seppä, K., Carbohydrate-deficient transferrin and conventional alcohol markers as indicators for brief intervention among heavy drinkers in primary health care. *Alcohol Clin. Exp. Res.*, 22, 892, 1998.
219. Sillanaukee, P. and Olsson, U., Improved diagnostic classification of alcohol abusers by combining carbohydrate-deficient transferrin and gamma-glutamyltransferase. *Clin. Chem.*, 47, 681, 2001.
220. Bell, H., Tallaksen, C., Sjåheim, T., Weberg, R., Räknerud, N., Örjasaeter, H., Try, K., and Haug, E., Serum carbohydrate-deficient transferrin as a marker of alcohol consumption in patients with chronic liver diseases. *Alcohol Clin. Exp. Res.*, 17, 246, 1993.
221. Gjerde, H., Johnsen, J., Bjørneboe, A., Bjørneboe, G.E., and Mørland, J., A comparison of serum carbohydrate-deficient transferrin with other biological markers of excessive drinking. *Scand. J. Clin. Lab. Invest.*, 48, 1, 1988.
222. Helander, A., Carlsson, A.V., and Borg, S., Longitudinal comparison of carbohydrate-deficient transferrin and gamma-glutamyl transferase: complementary markers of excessive alcohol consumption. *Alcohol Alcohol.*, 31, 101, 1996.
223. Randell, E., Diamandis, E.P., and Goldberg, D.M., Changes in serum carbohydrate-deficient transferrin and gammaglutamyl transferase after moderate wine consumption in healthy males. *J. Clin. Lab. Anal.*, 12, 92, 1998.
224. Niemelä, O., Sorvajärvi, K., Blake, J. E., and Israel, Y., Carbohydrate-deficient transferrin as a marker of alcohol abuse: relationship to alcohol consumption, severity of liver disease, and fibrogenesis. *Alcohol Clin. Exp. Res.*, 19, 1203, 1995.
225. Helander, A. and Carlsson, S., Carbohydrate-deficient transferrin and gamma-glutamyl transferase levels during disulfiram therapy. *Alcohol Clin. Exp. Res.*, 20, 1202, 1996.
226. Rosman, A.S., Basu, P., Galvin, K., and Lieber, C.S., Utility of carbohydrate-deficient transferrin as a marker of relapse in alcoholic patients. *Alcohol Clin. Exp. Res.*, 19, 611, 1995.
227. Keso, L. and Salaspuro, M., Laboratory tests in the follow-up of treated alcoholics: how often should testing be repeated? *Alcohol Alcohol.*, 25, 359, 1990.
228. Kristenson, H., Öhlin, H., Hulten-Nosslin, M.B., Trell, E., and Hood, B., Identification and intervention of heavy drinking in middle-aged men: results and follow-up of 24–60 months of long-term study with randomized controls. *Alcohol Clin. Exp. Res.*, 7, 203, 1983.
229. Kristenson, H. and Jeppsson, J.O., Drunken driver examinations. CD-transferrin is a valuable marker of alcohol consumption. *Lakartidningen*, 95, 1429, 1998.
230. Anton, R.F., Lieber, C., and Tabakoff, B., Carbohydrate-deficient transferrin and gamma-glutamyltransferase for the detection and monitoring of alcohol use: results from a multisite study. *Alcohol Clin. Exp. Res.*, 26, 1215, 2002.
231. Goodwin, D.W., Schulsinger, F., Hermanssen, L., Guze, S.H., and Windkure, G., Alcohol problems in adoptees raised apart from alcoholic biological parents. *Arch. Gen. Psychiatry*, 28, 238, 1973.
232. Cloninger, C.R., Bohman, M., and Sigvardsson, S., Inheritance of alcohol abuse: cross fostering analysis of adopted men. *Arch. Gen. Psychiatry*, 38, 861, 1981.
233. Cloninger, C.R., Neurogenetic adaptive mechanisms in alcoholism. *Science*, 236, 410, 1987.

234. Devor, E.J. and Cloninger, C.R., Genetics of alcoholism. *Annu. Rev. Genet.,* 23, 19, 1989.
235. Ball, D.M. and Murray, R.M., Genetics of alcohol misuse. *Br. Med. Bull.,* 50, 18, 1994.
236. Ratsma, J.E., Van Der Stelt, O., and Gunning, W.B., Neurochemical markers of alcoholism vulnerability in humans. *Alcohol Alcohol.,* 37, 522, 2002.
237. Cowen, M.S., Chen, F., and Lawrence, A.J., Neuropeptides: implications for alcoholism. *J. Neurochem.,* 89, 273, 2004.
238. von Knorring, A.L., Bohman, M., von Knorring, L., and Oreland, L., Platelet MAO activity as a biological marker in subgroups of alcoholism. *Acta Psychiat. Scand.,* 72, 51, 1985.
239. Begleiter, H., The collaborative study on the genetics of alcoholism. *Alcohol Health Res. World,* 19, 228, 1995.
240. Fowler, J.S., Volkow, N.D., Wang, G.-J., Pappas, N., Logan, J., MacGregor, R., Alexoff, D., Shea, C., Schlyer, D., Wolf, A. P., Warner, D., Zezulkova, I., and Cliento, R., Inhibition of monoamine oxidase B in the brains of smokers. *Nature,* 379, 733, 1996.
241. Agarwal, D.P., Molecular genetic aspects of alcohol metabolism and alcoholism. *Pharmacopsychiatry,* 30, 79, 1997.
242. Chen, C.C., Lu, R.B., Chen, Y.C., Wang, M.F., Chang, Y.C., Li, T.K., and Yin, S.J., Interaction between the functional polymorphisms of the alcohol-metabolism genes in protection against alcoholism. *Am. J. Hum. Genet.,* 65, 795, 1999.
243. Glucksman, E., Alcohol and accidents. *Br. Med. Bull.,* 50, 76, 1994.
244. Midanik, L., The validity of self-reported alcohol consumption and alcohol problems: a literature review. *Br. J. Addict.,* 77, 357, 1982.
245. Del Boca, F.K. and Darkes, J., The validity of self-reports of alcohol consumption: state of the science and challenges for research. *Addiction,* 98(Suppl. 2), 1, 2003.

CHAPTER 5

Alcohol Determination in Point of Collection Testing

J. Robert Zettl, B.S., M.P.A., DABFE
Forensic Consultants, Inc., Centennial, Colorado

CONTENTS

The chapter covers five topics: general considerations; pharmacology and toxicology of alcohol; organizational policies and procedures for specimen collection and testing; governmental regulations; and devices for testing of breath, saliva, and urine. A comprehensive discussion of alcohol pharmacology and toxicology and evidentiary breath testing can be found elsewhere. The material presented in those areas serves to assist the reader in understanding this chapter.

Devices used for human subject alcohol determination can be separated into four broad categories: (1) law enforcement — driving under the influence (DUI); (2) diagnostic for treatment or other medical purposes; (3) pre-employment and workplace for compliance; and (4) for cause and random for governmental compliance. This chapter focuses on devices used in the last two venues.

5.1 GENERAL CONSIDERATIONS

The primary focus of this chapter is alcohol point of collection test devices and procedures, but it is appropriate to discuss briefly how and why alcohol testing is important in point of collection testing.

According to information from the National Highway Traffic Safety Administration, there were 41,471 motor vehicle traffic fatalities in the U.S. in 2000.[1] Of those 41,471 fatalities, 15,935 or 38.4% were alcohol related. This represents an average of one alcohol-related fatality every 31 min.

The National Safety Council[2] estimates the economic loss to society from a single highway fatality to be $90,000, and the corresponding total economic loss exceeding $4 billion annually.[3] The drinking driver affects every one of us through increased taxes for additional law enforcement needs, medical facilities, incarceration, rehabilitation, social security and welfare for survivors, as well as increased insurance rates.

In the U.S., alcohol accounts for two thirds of all workplace substance abuse complaints and depletes a similar percentage from their health care benefit budgets. The results of a 2002 study[4] released by the Substance Abuse and Mental Health Services Administration (SAMHSA) showed drug use trends in the U.S. Of interest is that most alcohol and drug users are employed.

Alcohol abuse and its related problems cost society many billions of dollars each year.[5–8] Estimates of the economic costs of alcohol abuse attempt to assess in monetary terms the damage that results from the misuse of alcohol. These costs include expenditures on alcohol-related problems and opportunities that are lost because of alcohol. In a 1985 cost study, Rice and co-workers[9] estimated that the cost to society of alcohol abuse was $70.3 billion. By adjusting cost estimates for the effects of inflation and the growth of the population over time, that cost today could be well over $100 billion.

5.2 PHARMACOLOGY AND TOXICOLOGY OF ALCOHOL

Alcohol is commonly ingested orally and passes from the mouth, through the esophagus, into the stomach, and then into the small intestine. From here, alcohol is absorbed into the blood and distributed by the circulatory system to all parts of the body. As alcohol is transported through the body by the blood flow, it passes through the liver, which is primarily responsible for its metabolism, then to the kidneys where it is eliminated into the urine, then to the brain where it elicits its primary effect, and finally to the lungs where some alcohol passes unaltered out of the body. This unaltered alcohol permits the determination of a breath alcohol concentration (BrAC) from the alveolar or deep lung air.[10]

Alcohol is a low-molecular-weight organic molecule that is sufficiently similar to water to be miscible with water in all proportions. In addition, alcohol is able to cross cell membranes by a simple diffusion process; therefore, it can quickly achieve equilibrium throughout the body. The result of these properties is that alcohol rapidly becomes associated with all parts of the body, *including oral fluid,* and concentrations of alcohol will be found in proportion to body water content.

5.3 ORGANIZATIONAL POLICIES AND PROCEDURES

5.3.1 Collection and Testing

Although it is the quantity of alcohol present in the brain that actually affects a person's normal functions, practicality necessitates a specimen that is in equilibrium with the brain be used to reflect alcohol concentration.

Therefore, most studies center on the use of blood to correlate the degree of alcohol impairment; however, over the last 30-plus years, breath testing has supplemented blood as the specimen of choice. Due to the difficulty in the collection of urine, its use as a specimen has fallen into some disfavor. Blood is more likely to be used in defining driving under the influence of alcohol and blood and/or urine in defining driving under the influence of drugs.

Serum or plasma is often used in clinical situations where alcohol is tested. Since the water content of serum or plasma is greater than that of whole blood, serum/plasma alcohol concentrations are typically 10 to 20% greater than the corresponding blood specimen. Therefore, if serum/plasma tests are to be utilized where blood alcohol concentrations (BACs) define legal penalties, the serum/plasma concentrations must be corrected. Serum-to-blood ratios vary from 1.12 to 1.17 while plasma averages 1.18. A ratio of 1.16 is commonly used to make the conversion.[11]

Blood or urine is seldom used in alcohol point of collection testing (APOCT) because of difficulties that hinder its ease of collection. In most instances, existing regulations or statutes will dictate the choice of specimen. At present breath and saliva are the specimens collected in the majority of APOCT venues.[12]

5.3.2 Specimens for Analysis

The most commonly used specimen for APOCT is breath for the two venues covered in this chapter: (a) pre-employment and workplace for compliance and (2) for cause and random for governmental compliance. With recent technological advances, saliva is taking on a new dimension and is pressing breath as the "new" specimen of choice. Saliva can, with appropriate care in specimen procurement and application of the accepted distribution ratios, be correlated with blood and a blood alcohol equivalent can be reported.

5.4 BREATH TESTING FOR ALCOHOL

5.4.1 Principles

Breath tests to determine the alcohol concentration present in a person's body are by far the most frequently utilized tests in cases involving driving under the influence of alcohol. States today have universally adopted legislation that permits reporting of a subject's alcohol concentration in breath units of alcohol per 210 L of breath. Breath alcohol analysis is the method of choice of law enforcement and many others due to ease and operational simplicity of new generation breath testing equipment, speed with which analyses may be conducted, convenience of being able to perform the analysis at or near the scene of an incident, and the convenience of having the subjects' test results immediately available.

5.4.2 Breath Alcohol Testing Devices

5.4.2.1 Introduction

The National Highway Traffic Administration has established a conforming products list for instruments that conform to the "Model Specifications for Devices to Measure Breath Alcohol."[13] This list contains all devices currently approved to perform breath alcohol testing within the U.S. Although many of the devices found on the present conforming products list are no longer manufactured, many are still in use.

In 1995, Zettl in conjunction with the Colorado Department of Public Health and Environment conducted a national survey of alcohol programs.[14] Some "electronic" devices listed in the 1995 Colorado Department of Public Health survey are no longer manufactured and newer-generation devices have been introduced since that survey was completed. Refer to the National Highway Traffic Administration conforming products list for a complete listing of all "electronic" instruments that can be used for point of collection testing.

5.4.2.2 Electronic Devices — Evidentiary

Electronic devices are generally classified into two distinct categories: (1) tabletop devices, which are larger and more expensive units originally designed for use in DUI testing, and (2) handheld or preliminary breath (alcohol) testers (PBTs), which are smaller units originally designed for use as screening devices by law enforcement to determine a suspected DUI's approximate BrAC level at the time of stop. PBTs are now used extensively as evidentiary units in workplace and DUI testing.

Both types of devices are designed to analyze a breath sample and determine the amount of alcohol present in such a manner that the results have a degree of scientific accuracy and specificity sufficient to be reliable for presentation in court as evidence. They are self-contained portable laboratories in which the underlying principle, mode of operation, and safeguards are such that the end user can effectively operate the instruments and develop reproducible results.

5.4.2.3 Electronic Tabletop Devices

Infrared (IR) technology utilizes the principle that alcohol present as a vapor in breath absorbs specific wavelengths of IR light. Alveolar air is trapped in a sample chamber. IR light is directed through the sample cell and finally reaches a detector that measures the amount of light absorbed. As the concentration of alcohol vapor increases in the chamber, the amount of IR energy reaching the detector falls in a predictable exponential manner; hence, IR devices measure alcohol by detecting the decrease in the intensity of IR energy as it passes through the chamber.

The Draeger Corporation, Breathalyzer Division (Durango, CO) and Intoximeters, Inc. (St. Louis, MO) have developed breath test instruments that incorporate IR technology combined with an electrochemical (EC) fuel cell. Use of dual technology enhances both the quality and integrity of the sample and the accuracy of the alcohol test result. Fuel cell technology is discussed in Section 5.4.2.5.

The most recent device, the Intoxilyzer 8000, uses dual IR (two separate wavelengths), one for alcohol concentration and the other for interferent detection.

5.4.2.4 Electronic Gas Chromatographic Tabletop Devices

These devices are no longer being manufactured but may still be in use. They are the Gas Chromatograph Intoximeter (GCI) Mark II and Mark IV manufactured by Intoximeters, Inc. and the Alco-Analyzer Gas Chromatograph, models 1000, 2000, and 2100, manufactured by Luckey Laboratories and later US Alcohol Testing.

Although the instruments used different methods of detection, they were of equal specificity and accuracy. They could be used as either direct measurement devices or they could analyze collected samples such as blood, delayed breath, saliva, and urine by means of headspace chromatography. They were excellent for use in a centralized facility setting; that is, samples could be collected in the field and forwarded to a centralized processing facility where technical personnel conducted the testing.

5.4.2.5 Electronic Handheld Devices: PBTs

Handheld "electronic screening devices" are small, portable, relatively inexpensive, may cost only a few hundred dollars, and were originally designed to estimate the BAC or BrAC of an individual. Some of the devices may be prone to erroneous readings both falsely high and low. The electrical sensor instruments — fuel cell — are generally more accurate (±10% range or better) and some are more specific for alcohol than others with printer attachments to be used in evidential settings. As stated, they were originally intended to assist in quickly determining the approximate alcohol concentration in an individual and were referred to as PBTs. The devices are useful for monitoring and controlling alcohol abuse and/or intoxication in various workplaces, alcohol abuse programs, and correctional institutions.

Evidentiary handhelds utilize an EC sensor — generally a fuel cell — to measure the amount of alcohol present in the sample. All of the devices in this category use alveolar air and may be calibrated for BrAC concentration.

Because the devices were originally intended for screening, they should be used carefully and only for their intended purpose by trained personnel. Some of the older-generation devices are not sufficiently accurate for evidentiary purposes and if the subject tested is placed at risk pending the results of such tests, then the initial "screening" test should be confirmed by an evidentiary test. PBTs are extremely useful because they are generally less expensive than tabletop units and training and upkeep are less involved. Newer-generation PBTs, when used according to manufacturer specifications, can yield an accurate BrAC.

5.4.2.6 Pass–Warn–Fail Devices

These devices are calibrated to determine into which of three likely broad-range categories of BrAC an individual will likely fall.

PASS — Alcohol concentration below a predetermined level. Usually a level at which the individual is considered able to drive safely. Usually less than 0.05% except for underage drinkers.

WARN — Alcohol concentration above the level where a person would pass but below a level where the individual is considered intoxicated. Usually at or above 0.05% but less than the state's legal limit per se.

FAIL — Alcohol concentration above the level where the individual is considered intoxicated. Usually at or above 0.08% or 0.10%. Depends on the state's level for DUI.

The concentrations at each level may be arbitrarily set as desired. Refer to Table 5.1 for a listing of some of the Pass–Warn–Fail devices and Table 5.2 for a listing of some of the digital display devices.

Table 5.1 Pass–Warn–Fail PBTs

1. B.A.T. III by Century Systems, Inc. uses catalytic combustion for alcohol analysis. Pointer indicates Warn or Fail.
2. Alcohalt Detector by Mine Safety Appliance Company uses catalytic combustion for alcohol analysis and two indicator lights for Pass or Fail.
3. A.L.E.R.T. Model J3AD by Alcohol Counter Measures, Inc. uses a Taguchi semiconductor detector to analyze the alcohol and a series of green, amber, and red lights to indicate Pass, Warn, or Fail.
4. Older generation Alco-Sensor by Intoximeters, Inc. used a cluster of light-emitting diodes to indicate Pass, Warn, Fail (later models use a digital display for a direct readout of the % alcohol present).

Table 5.2 Digital Display PBTs

1. Alco-Sensors II through IV by Intoximeters, Inc. The Alco Sensor IV is shown in Figure 5.1.
2. Phoenix by Life Loc. See Figure 5.2.
3. CMI Corporations S-D2 and Models 200, 300, and 400. See Figure 5.3 for the S-D2.
4. National Draeger's Alcotest 7410. See Figure 5.4.

5.4.2.7 Digital Display Devices

There are many digital display devices on the NHTSA conforming products list. Some are shown in Table 5.2 (Figures 5.1 through 5.4). All use a fuel cell to determine the amount of alcohol present and %BrAC is digitally displayed.

5.4.2.8 Handheld — Non-Evidentiary

According to the manufacturer's product information the Alco-Scan Model AL-2500 (Figure 5.5) is a very versatile device that measures %BAC. The user gently blows at the sensor's intake and within 2 s the LCD displays a BrAC. The AL2500 is a compact and ideal device for use in social gatherings and for group testing.[15]

There are many devices like the Alco-Scan Model AL-2500, many priced under $100, but the end user should be cautioned that many are non-evidentiary and may not be suitable for some testing venues.

5.4.2.9 Screening Devices

NHTSA establishes which devices can be used as screening devices under its "Conforming Products List of Screening Devices to Measure Alcohol in Body Fluids."[16]

5.4.3 Saliva-Based Technology

5.4.3.1 Analytical Principle

Saliva collection for alcohol testing is regularly employed in POCT facilities but is not used in DUI testing due to its practical constraints in court. Although saliva may be impractical for DUI enforcement because significant subject cooperation is needed to facilitate collection, recent advances in collection technology hold great promise for its use in POCT on-site (roadside) drug detection. For additional information, visit the ROadSIdeTestingAssessment Web site.[17]

Saliva alcohol results can be compared to the amount of alcohol contained in a person's blood. If collected properly by observing a waiting period after a person has consumed his or her last alcoholic beverage, usually 10 to 15 min, then any residual alcohol will have been absorbed, swallowed, or evaporated, and the person's mouth is "clear." According to one manufacturer's information the relationship between the amount of saliva alcohol and blood alcohol is 1:1, whereas with breath it is 000048:1, making saliva a more sensitive testing medium than breath.[18]

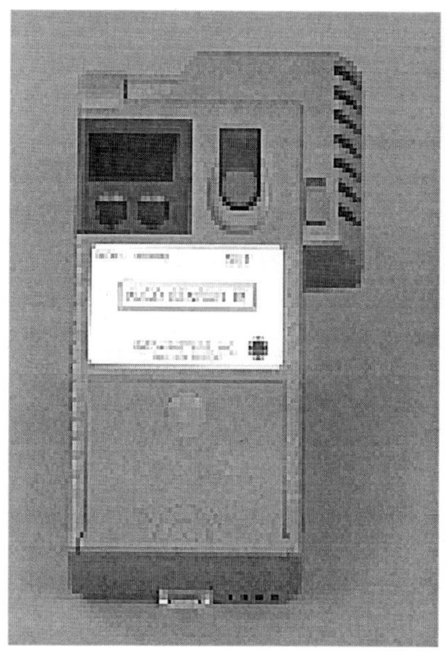

Figure 5.1 Intoximeters, Alco Sensor IV.

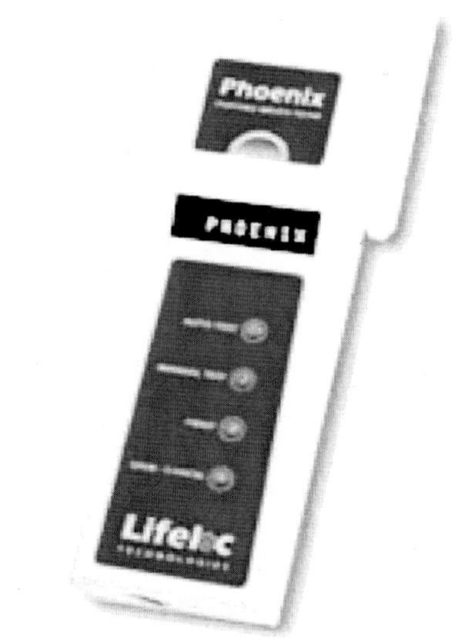

Figure 5.2 Lifeloc, Phoenix.

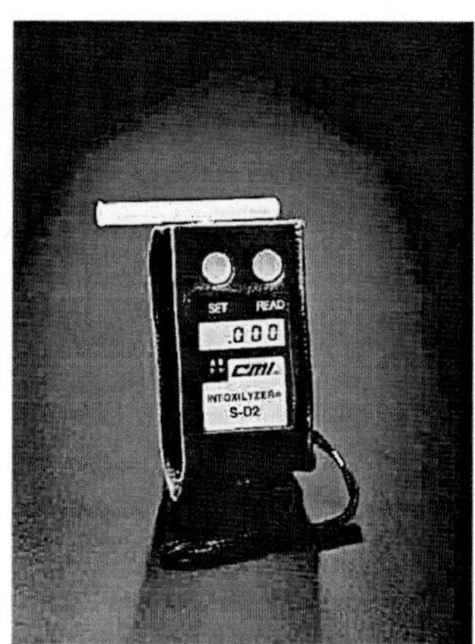

Figure 5.3 CMI SD-2.

Figure 5.4 Draeger 7410.

Figure 5.5 Alco-Scan Model AL-2500.

There are two prominent saliva-based alcohol test procedures. The QED saliva alcohol test procedure will not react with ketones often found in the saliva of patients with diabetes. Unlike some breath analyzers and other saliva tests, the QED is specific to ethyl alcohol and will not cross-react with acetone and ketones produced by diabetics.

The second type of disposable tester, strip test technology, does not have a great correlation between a person's true BAC and saliva alcohol concentration. Strip-based saliva testers are treated with an enzyme alcohol oxidase, which responds to alcohol in proportion to the concentration of alcohol in a mixed saliva sample placed on it. The user estimates the BAC by comparing the color change on the test strip patch to standard colors calibrated to correspond to different BACs. Although some saliva testers seem to indicate the presence of alcohol well, the enzyme alcohol oxidize used in these testers is easily affected by hot and cold temperatures. Hot temperatures will tend to indicate falsely high readings, while cold temperatures will tend to indicate falsely low readings.

Exposure to temperatures above 80°F or to ambient air will destroy the enzyme alcohol oxidase, rendering the tester useless. Most saliva testers give no indication if contamination has occurred, and if it has, they may not work effectively. Saliva testers generally have a shelf life of 1 year or less.

The technology and chemical reaction employed in the QED or the test strip technology is not as precise, accurate, or reliable as breath alcohol testing. Saliva-based alcohol tests require an evidential breath test (EBT) to confirm positive test results. Saliva alcohol testing is much less expensive to operate than a breath test, and unless a POCT facility conducts a very high volume of tests in a central location, then saliva testing instead of breath may be more cost-effective. Since most employees do not test positive for alcohol, simple screening is generally more cost-effective for POCT facilities.

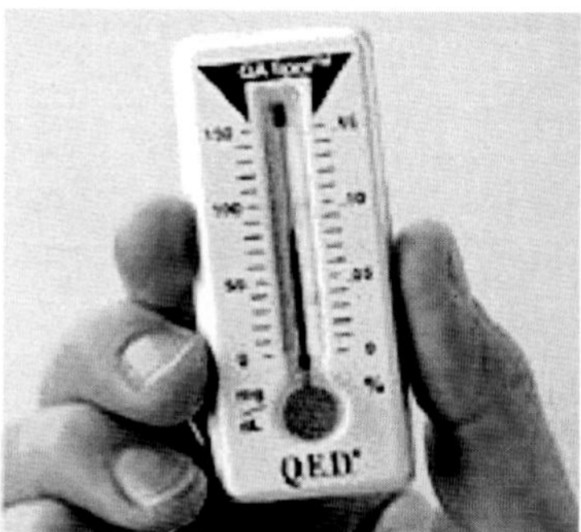

Figure 5.6 QED Saliva Alcohol Test.

5.4.3.2 Saliva-Based Devices

QED Saliva Alcohol Test — The QED (Figure 5.6) is a quantitative test device for the rapid determination of equivalent BAC using a non-invasive saliva sample. Approved by the Federal Department of Transportation (DOT) for commercial alcohol testing programs, the QED uses a unique patented lateral flow method to rapidly determine alcohol presence in saliva expressed as %BAC and milliliter per deciliter concentration. It is as simple as reading a thermometer.[19]

The QED Saliva Alcohol Test uses a preset chemical reactive process that requires no user intervention; a color bar rises to the level of alcohol present in the system in much the same way as a mercury thermometer. In extensive clinical trials, saliva alcohol levels measured by the QED Saliva Alcohol Test demonstrated a high correlation rate of 98% ($r = 0.98$) to blood analyzed by sophisticated laboratory gas chromatography methods.[20]

The QED Saliva Alcohol Test (Figure 5.6) is an easy-to-use diagnostic procedure with everything required contained in a sealed foil package. Total time required for the test is between 3 and 5 min. The three basic steps are as follows:

1. Using the cotton swab included, actively swab around the cheeks, gums, and tongue for 30 to 60 s or until the cotton swab is completely saturated with saliva.
2. Place the test device on a flat surface. Gently twist the swab with the collected saliva sample into the entry port and apply steady pressure to activate the capillary action until the pink fluid passes the QA Spot™ located at the top of the test device.
3. Allow the test device to develop for 2 min. A distinct purple bar will form within the marked scale region. The highest point of the purple bar represents the level of alcohol expressed either as a percentage or as grams per 100 ml or milligrams per deciliter concentration.

According to product information, the QED Saliva Alcohol Test will accurately measure a range of BAC of 0 to 145 mg/dl or 0.000% to 0.145% equivalent BAC.[20]

POCT facilities using the saliva alcohol test in very remote areas can comply with the DOT requirement that confirmation tests on positive screening tests must be conducted within 30 min. DOT will accept results of confirmation tests conducted more than 30 min after a positive screening test (49 CFR Part 40 section 40.65, paragraph (b)).[20] The DOT added a sentence, which directs the Breath Alcohol Technician (BAT) to simply explain "why?" if a confirmation test is done more than 30 min after a screening test.

Saliva-Based Test Strip

Alco-Screen: The ALCO-Screen™ (Figure 5.7) saliva alcohol test is intended for use as a rapid, highly sensitive method to detect the presence of alcohol in saliva and to provide a semi-

Figure 5.7 ALCO-Screen.

quantitative approximation of BAC. For applications where a quantitative determination of BAC is required, a positive ALCO-Screen result must be verified using an acceptable quantitative alcohol analysis procedure. ALCO-Screen requires no special training provided instructions are followed carefully. However, a qualified professional should perform quantitative follow-up testing. ALCO-Screen is not intended as a measurement of mental or physical impairment but rather a screening test for the presence of alcohol in semiquantitative amounts. As with any saliva-based or breath alcohol tester, a deprivation period of at least 15 min must be observed before beginning the test. This includes non-alcoholic drinks, tobacco products, coffee, breath mints, food, etc. The ALCO-Screen is used by saturating a reactive pad with saliva from the test individual's mouth or sputum cup. At exactly 2 min, a change in color is observed in the reactive pad. A color change of green or blue indicates the presence of alcohol and a positive result. Results obtained after more than 2 min and 30 s (2.5 min) may be erroneous and should not be used. A BAC is estimated by comparing the color of the reactive pad to the color chart on the back of the test package (Figure 5.7). The ALCO-Screen produces a color change in the presence of saliva alcohol ranging from a light green-gray color at 0.02% BAC to a dark blue-gray color near 0.30% BAC. ALCO-Screen is designed and calibrated to be interpreted 2 min after saturation of the reactive pad. Waiting longer than 2 min to interpret the test can result in erroneous or false-positive results. ALCO-Screen is a visually interpreted test; as such, exact interpretation of results is not required in most cases. However, persons who are color blind or visually impaired may experience difficulty when a more specific interpretation is required. Furthermore, where test interpretation may be biased for whatever reason, it is suggested that another person's opinion of test results or color matching be obtained.[20]

ALCO Screen 2: ALCO-Screen 2 is a simple and cost-effective method of monitoring for alcohol consumption in a zero tolerance testing program. According to its product information, the ALCO-Screen 2 has been tested and approved by the U.S. DOT for required testing of all transportation and safety-sensitive employees for BACs above the federally mandated zero tolerance level of 0.02% (Figure 5.8). ALCO-Screen 2 is a simple one-step saliva-screening test that works in a clean, non-invasive manner and provides results in 4 min. Simply wet the test pad with saliva and wait 4 min. The development of a line on the test pad at 4 min indicates a BAC exceeding 0.02%. Any line, no matter how faint, developing on the reactive test pad at 4 min is a positive

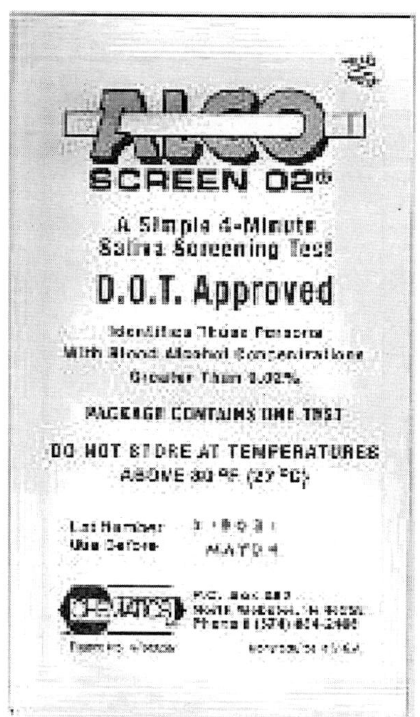

Figure 5.8 ALCO-Screen 2.

test. The Alco-Screen 2 is highly sensitive and can be used for evidentiary purposes. Completed test results can be photocopied for permanent filing.[20]

5.4.4 Chemical–Color-Change-Based Devices

Subcategories of screening devices, which are not electronic, make a determination of alcohol concentration by use of a chemical reaction. The first type of non-electronic device consists of either dichromate or permanganate salts in acid-impregnated crystals, which are placed in glass tubes. The individual being tested blows into a balloon or plastic bag or through the tube. After a certain volume of air or time has transpired, a measurement of the length of stain on the crystals in the tube (color change) is used to approximate the BrAC. The color change is a result of the chemical reaction occurring between alcohol and the chromate or permanganate salts in the crystals. Examples of older devices of this type include the Alcolyzer, several varieties distributed by Intoximeters, Inc., the Becton-Dickinson devices, Kitigawa Drunk-O-Tester by the Komo Chemical Industrial Company, Sober-Meters (Mobats) by US Alcohol, and AlcoPro (Knoxville, TN). These screening devices use a mixed expired breath sample with the exception of the Becton-Dickinson device, which uses a two-chambered plastic bag to obtain alveolar air for the screening test. The results obtained from using these devices should be read according to time requirements expressed by the manufacturer. Other oxidizable components of breath will continue to react with the chemicals and may produce false positives.

Screening devices that utilize a color change reaction for alcohol detection are disposable and good for only one test, whereas electronic devices have an extended life and can be used repeatedly after resetting; hence they may be more cost-effective if an agency is doing multiple testing.

One of the more popular disposable screeners is the BreathScan® Alcohol Detector (Figure 5.9).[20] According to the manufacturer's promotional material, "it is a disposable breath-alcohol indicator designed for one-time use" and according to its manufacturer it provides an accurate measure of the alcohol present in the exhaled breath of a test subject. By measuring the alcohol

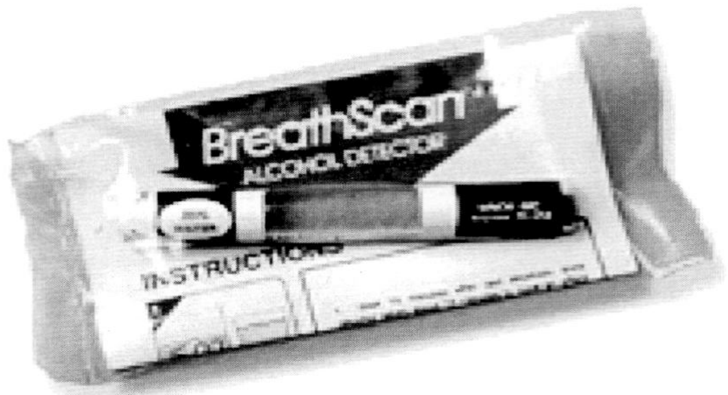

Figure 5.9 BreathScan Alcohol Detector.

content in the breath, a reliable indication of the blood alcohol level is achieved. The BreathScan detector employs a new, patented technology for simple, on-the-spot screening for the presence of alcohol. The BreathScan tester can only be used once and then disposed of, minimizing contamination associated with repeated use of nondisposable units (no AIDS cross transmission).

The BreathScan's low cost and ease of use make the tester ideal for screening to determine whether an individual should submit to a forensic-quality blood test for confirmation. Just break the internal capsule, shake, and blow hard into the test cylinder of breath alcohol detector for a few seconds. Then read the color change of the chemical crystals in 2 min or less. Approved by the DOT, the detectors are available in five BAC levels for a complete range of sensitivity: 0.10%, 0.08%, 0.05%, 0.04%, and 0.02% (for zero tolerance testing), and they are very light and easy to carry around, weighing 0.16 oz each.

5.4.5 Passive Alcohol Sensor Devices

"Passive" alcohol sensing (PAS) devices are designed to detect the presence of alcohol in a person's normally expelled breath; that is, the subject being tested is not required to blow into a mouthpiece as with conventional breath test devices. Passive alcohol sensing devices pull, through the use of a fan or other mechanical means, the vapor from the subject's normal breathing when the device is activated and held in close proximity to the subject's mouth. The device can also be held over open containers of an alcohol beverage to see if an underage person is drinking illegally. The present distributor of the PAS Systems is LLC (Fredericksburg, VA). It markets the PAS III, non-invasive alcohol-screening instrument, which has a built-in high-intensity flashlight (Figure 5.10). Their product information refers to the device as a "sniffer" for overt or covert alcohol detection. This device uses fuel cell technology for determining alcohol concentration. Another PAS device, the Alcometer, is currently available from Lions Laboratories (Cardiff, Wales, U.K.).

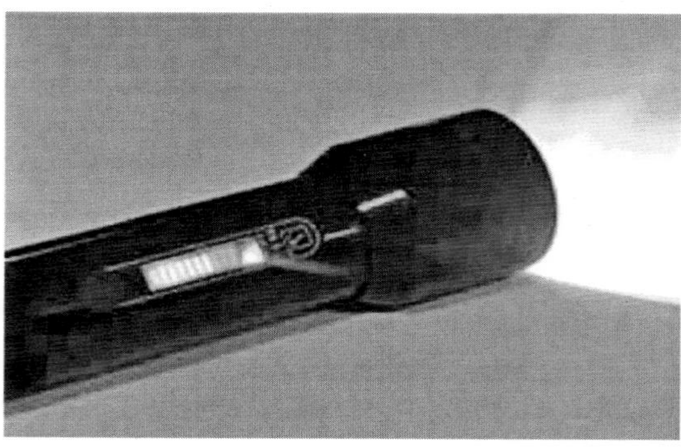

Figure 5.10 PAS IV "sniffer."

Passive alcohol sensors have had a varied history — first introduced in the early 1970s without much success. In a recent DOT/NHTSA study (DOT HS 807 394), one such device was able to discriminate among differing alcohol air samples under laboratory conditions. The user has to be cognizant that passive alcohol sensors are influenced by wind disturbances. Wind or any air movement tends to invalidate their proficiency.

5.5 BLOOD AND URINE — COLLECTION, IDENTIFICATION, AND PRESERVATION OF SPECIMENS

This issue is addressed only briefly here. Other treatises and handbooks can be found that will assist the reader in understanding this area of alcohol testing.[21,22] Blood or urine specimens must be collected in a manner to maintain the chain of custody as in any other forensic case. However, additional precautions are required since the specimens are biological in nature, namely, removal of blood by qualified medical persons in an alcohol-free manner, and preservation of the specimen to permit mailing and long-term storage. Before establishing a system for collection, one should consult with the certifying state or other agency in control of specimen collection and testing to prevent unnecessary problems.

5.6 QUALITY ASSURANCE AND PROFICIENCY TESTING

5.6.1 Saliva

Quality control (QC) requirements for the saliva alcohol test can be conducted using control checks, with the Saliva Alcohol Ethanol Control from OraSure Technologies.[23] Saliva alcohol ethanol control solutions should be run once per lot number of saliva alcohol tests.

Saliva-based devices such as the ALCO-Screen may be qualitatively verified using a test solution prepared by adding 4 drops of 80-proof distilled spirits to 8 oz (1 glass) of water. This solution should provide a color reaction equal to or higher (darker) than the 0.04% color block. The color reaction with alcohol in saliva is somewhat slower and less intense than with alcohol in aqueous solutions.

5.6.2 Body Fluids

A laboratory conducting blood alcohol determinations whether for clinical or forensic purposes should maintain an internal system designed to assure the reliability of all laboratory data and should participate in an external proficiency testing program, where available, that evaluates the laboratory on the basis of the comparability of its results with those of several reference laboratories analyzing the same sample.

The quality assurance program should include maintenance and periodic testing of equipment, validation and recalibration of methods, reagent evaluation, and surveillance of results. In-house reference calibrators and controls may be prepared from outdated whole human blood targeting concentrations of 0.000, 0.050, 0.100, 0.200, and 0.400% ethanol. Various reference materials are available to prepare or serve as standards, calibrators, and controls, e.g., the National Institute of Standards and Testing (NIST) material, SRM 1821 Ethanol (formerly National Bureau of Standards), and the College of American Pathologists (CAP) alcohol reference materials. Many states, private entities, and reagent manufacturers provide reference, calibrator, and control materials.

For standardization, calibrators are assayed in triplicate and an appropriate standard curve prepared. Standardization should be repeated periodically or as dictated by changes in operational protocol. One or more quality control blood specimens (0.080, 0.015%, etc.) should be prepared and the mean and standard deviation determined for a total of 20 samples analyzed over a period of 10 days. The quality control sample should then be analyzed with every run of unknown alcohol samples and the result should fall within 95% confidence limits. The American Academy of Forensic Sciences, Toxicology Section and the Society of Forensic Toxicologists, Inc. have approved a quality assurance program titled, "Forensic Toxicology Laboratory Guidelines."[24]

These same quality control procedures should be used for the testing of any other body fluids such as urine, serum, saliva, and post-mortem samples. An excellent resource for quality assurance can be found in Garriott's *Medicolegal Aspects of Alcohol*[22] or by obtaining the "Forensic Toxicology Laboratory Guidelines" from the American Academy of Forensic Sciences, Toxicology Section or the Society of Forensic Toxicologists, Inc.[24]

Duplicate aliquoting and testing of forensic biological specimens are important parts of any quality assurance procedure or program. In general, the results of the two independent tests should fall within a range of each other by ±10% or 0.02% BAC. Other accuracy or precision criteria may be used, but an increased degree of confidence in the reported results is achieved by duplicate testing.[25]

5.6.3 Breath

Although correlation between alcohol concentrations calculated using breath testing instruments and alcohol concentrations determined directly from blood has been well documented in the literature and accepted by the courts, it is necessary to have a method for the standardization and quality control of breath test devices. For scientific and legal reasons, it is necessary to demonstrate that a particular device was functioning properly at the time a subject was tested.

5.6.3.1 Wet Bath Breath Alcohol Simulation

The relationship between the concentration of alcohol in air as compared to blood at 34°C is discussed elsewhere. The partition ratio for air/blood is greater than that of air/water; therefore, if an aqueous alcohol solution is heated to 34°C the amount of alcohol in the air space at equilibrium will be less than that of blood. To produce a breath sample that simulates a given BAC, it is therefore necessary that the aqueous (water and alcohol) solution be of an alcohol concentration greater than the expected reading value. If an expected BrAC of 0.100% at 34°C is required, then the aqueous water and alcohol concentration must be 0.121%.

5.6.3.2 Simulation Units for Breath Alcohol Testers

There are several manufactures of breath alcohol simulation devices. A complete list of companies and their devices can be obtained by contacting the National Highway Traffic Safety Administration, Office of Alcohol and State Programs, 400 Seventh Street, S.W., Washington, D.C. 20590, (202) 366-5593 and asking for the current Model Specifications for Calibrating Units for Breath Alcohol Testers.

The simulator maintains the temperature at a constant 34 ± 0.2°C. The simulator contains several basic components: the first is the jar that holds the solution; the second is the head, which serves as a seal to the jar and contains the thermostat, thermometer, propeller, and motor, an air inlet tube that is attached to a bubbler tube, an air outlet tube, and a wire mesh baffle to prevent the solution from escaping through the outlet tube.

The simulator solution is the most critical component. It should be carefully prepared and its accuracy checked by a competent laboratory and standardized by chemical and chromatographic analysis, comparing the results to a primary reference standard material (e.g., potassium dichromate, NIST Alcohol Standard).

5.6.3.3 Gas Breath Alcohol Simulation

Breath alcohol simulation is achieved by using a dry gas mixture. Several manufacturers prepare or sell gaseous ethanol products. The device consists of a tank of pressurized air containing a specified alcohol concentration; a button is depressed and the alcohol/air mixture is released into the breath-testing device. These gaseous standards are useful if the breath test instrument utilizes a relatively small sample volume. For breath-testing instruments with large sample volumes, use of these gaseous standards may or may not be practical. Any pressurized gas mixture of alcohol and air is subject to variation due to atmospheric pressure; hence, the mixture should be standardized against a wet solution of known alcohol concentration prior to use in a field situation.

For a complete listing of the dry gas ethanol manufacturers, contact DOT/NHTSA and obtain its most recent listing of these units, contained in the conforming products list of calibrating units.[26]

5.7 CONCLUDING REMARKS

The abbreviation BAC refers to blood alcohol concentration, with concentrations expressed as percent weight to volume, % (w/v) or grams of alcohol per 100 ml of blood. The abbreviation BrAC refers to breath alcohol concentration expressed as percent weight to volume, % (w/v) or grams of alcohol per 210 L of deep lung or alveolar breath. The term *alcohol* refers to ethyl alcohol or ethanol. There are many excellent resources for forensic alcohol information; those by Garriott and Saferstein are especially helpful. In addition to printed documentation, numerous Internet sites are available. Table 5.3 provides a list of some of these.

ACKNOWLEDGMENTS

The author thanks Yale Caplan, Ph.D., and the manufacturers of alcohol test equipment for providing assistance, information, and photographs.

Table 5.3 Alcohol and Traffic Safety-Related Sites on the Internet

AAA Foundation for Traffic Safety	www.aaafts.org
Air Products	www.airproducts.com
Alcohol Countermeasure Systems	www.acs-corp.com
American Academy of Forensic Sciences	www.aafs.org
Bureau of Transportation Statistics	www.bts.gov
Canadian Safety Council	www.safety-council.org
CMI, Inc.	www.alcoholtest.com
Draeger Breathalyzer Division	www.drager-breathalyzer.com
Drug and Alcohol Testing Industry Association	www.datia.com
Guth Laboratories	www.guthlabs.com
Health and Human Services Drug Testing	www.health.org/workpl.htm
Insurance Institute for Highway Safety	www.carsafety.org
International Association for Chemical Testing	www.iactonline.org
International Council on Alcohol, Drugs and Traffic Safety	raru.adelaide.edu.au/icadts/
Intoximeters, Inc.	www.intox.com
Lifeloc	www.lifeloc.com
Lion Laboratories	www.lionlaboratories.com
Mothers Against Drunk Driving	www.madd.org
National Clearinghouse for Alcohol and Drug Information	www.health.org
National Committee for Clinical Laboratory Standards	www.nccls.org
National Highway Traffic Safety Administration	www.nhtsa.dot.gov
National Institutes of Health	www.nih.gov
National Institute on Alcohol Abuse and Alcoholism	www.niaaa.nih.gov
National Institute on Drug Abuse	www.nida.nih.gov
National Motorists Association	www.motorists.org
National Safety Council, Committee on Alcohol & Other Drugs	www.nsc.org
NPAS DataMaster	www.npas.com
PAS Systems International	www.sniffalcohol.com
Road Side Testing Assessment	www.rosita.org
Scott Specialty Gasses	www.scottgas.com
Society of Forensic Toxicologists	www.soft-tox.org
Substance Abuse and Mental Health Services Administration	www.samhsa.gov
Transportation Research Board	www.nas.edu/trb
U.S. Department of Health and Human Services	www.hhs.gov
U.S. Department of Transportation	www.dot.gov

REFERENCES

1. January 2000 Impaired Driving Program Update, National Highway Traffic Safety Administration, Traffic Safety Programs, Impaired Driving Division, Washington, D.C., 2000.
2. National Safety Council, 1121 Spring Lake Drive, Itasca, IL.
3. Colorado Association of Chiefs of Police, D.U.I. Enforcement Manual for the State of Colorado, August 1977.
4. Results from the 2002 National Survey on Drug Use and Health: National Findings. Department of Health and Human Services, Substance Abuse and Mental Health Services Administration, Office of Applied Studies, Washington, D.C., 2002.
5. Berry, R.E., Boland, J.P., Smart, C., and Kanak, J., *The Economic Cost of Alcohol Abuse: 1975,* Policy Analysis, Brookline, MA, 1977.
6. Cruze, A.M., Harwood, H.J., Kristiansen, P.L., Collins, J.J., and Jones, D.C., *Economic Costs to Society of Alcohol and Drug Abuse and Mental Illness: 1977,* Research Triangle Institute, Research Triangle Park, NC, 1981.
7. Harwood, H.J., Napolitano, D.M., Kristiansen, P.L., and Collins, J.J., *Economic Costs to Society of Alcohol and Drug Abuse and Mental Illness: 1980,* Research Triangle Institute, Research Triangle Park, NC, 1984.

8. Rice, D.P., Estimating the Cost of Illness. Health Economics Series, No. 6. DHEW Pub. No. (PHS) 947-6, 1966, U.S. Department of Health, Education and Welfare, Rockville, MD, 1966.
9. Rice, D.P., Kelman, S., Miller, L.S., and Dunmeyer, S., *The Economic Costs of Alcohol and Drug Abuse and Mental Illness: 1985,* National Institute on Drug Abuse, Rockville, MD, 1990.
10. Zettl, J.R., Prosecution of driving while under the influence student manual, in Toxicology and the Forensic Analysis of Alcohol, American Prosecutors Research Institute, National Traffic Law Center, Washington, D.C., USDOT/NHTSA Project Number 004NTLC and 0922 Drug Driver.
11. Payne, J.P., Hill, D.W., and King, N.W., Observations on the distribution of alcohol in blood, breath and urine, *Br. Med. J.*, 1, 196, 1996.
12. Dubowski, K.M. and Caplan, Y.H., Alcohol testing in the workplace, in *Medicolegal Aspects of Alcohol*, 3rd ed., J.C. Garriott, Ed., Lawyers & Judges, Tucson, 1996, 439–475.
13. National Highway Traffic Safety Administration, Highway Safety Programs, Model specifications for devices to measure breath alcohol, amended, *Fed. Regis.*, 67(192), 2002.
14. Zettl, J.R., *Colorado Alcohol Test Program Survey — Update*, Colorado Department of Public Health and Environment, Laboratory and Radiation Services Division, February 27, 1997.
15. Advance Safety Devices, 21000 Osborne Street, Suite 4, Canoga Park, CA 91304.
16. National Highway Traffic Safety Administration, Highway Safety Programs, Conforming products list for screening devices to measure alcohol in body fluids, amended, *Fed. Regis.*, 66(87), 2001.
17. Roadside Testing Assessment: www.rosita.org//.
18. Craig Medical Distribution, Inc., 185 Park Center Drive, Building P, Vista, CA 92801.
19. OraSure Technologies, Inc., Bethlehem, PA (formerly STC Technologies, Inc.).
20. STC Technologies, Inc., 1745 Eaton Avenue, Bethlehem, PA 18018-1799.
21. Caplan, Y.A. and Zettl, J.R., The determination of alcohol in blood and breath, in *Forensic Science Handbook,* Vol. 1, 2nd ed., R.E. Saferstein, Ed., Prentice Hall, Upper Saddle River, NJ, 2001, chap. 12.
22. Garriott, J.C., Ed., *Medicolegal Aspects of Alcohol*, 3rd ed., Lawyers & Judges, Tucson, 1996.
23. OraSure Technologies, Inc., 220 East First Street, Bethlehem, PA 18015.
24. *Forensic Toxicology Laboratory Guidelines,* American Academy of Forensic Sciences, Toxicology Section and Society of Forensic Toxicologists, Inc., 1997–1998.
25. Jones, A.W. and Logan, B.K., DUI defenses, in *Drug Abuse Handbook*, S. Karch, Ed., CRC Press, Boca Raton, FL, 1998, 1006–1045.
26. Department of Transportation, National Highway Traffic Safety Administration, Model specifications for calibrating units for breath alcohol testers; conforming products list of calibrating units, *Fed. Regis.,* 62(156), August 13, 1997.

Index

A

B

C

D

E

F

G

H

I

J

K

L

Q

R

S

T

U

V

W

Y

Z